Th
wo
and

rec
act
in
end
me
tra
grc

po
ass
cor
cre
cie

the
wo
ho

Ro
Un
Po

Housing Women

Edited by
Rose Gilroy and Roberta Woods

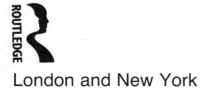

London and New York

First published 1994
by Routledge
11 New Fetter Lane, London EC4P 4EE

Simultaneously published in the USA and Canada
by Routledge
29 West 35th Street, New York, NY 10001

© 1994 Selection and editorial matter, Rose Gilroy and
Roberta Woods; individual chapters, the contributors

Typeset in Times by
Florencetype Ltd, Stoodleigh, Devon
Printed and bound in Great Britain by
Biddles Ltd, Guildford and King's Lynn

British Library Cataloguing in Publication Data
A catalogue record for this book is available from the
British Library

Library of Congress Cataloging in Publication Data
A catalogue record for this book has been requested

ISBN 0–415–09462–3 (hbk)
ISBN 0–415–09463–1 (pbk)

Contents

Illustrations

Contributors

Marion Brion, Ph.D., FCIH, Cert. Ed. is now a freelance trainer, researcher and psychotherapist. In 1974 she initiated the first national research into housing education and training. She went on to develop and teach qualification and skills-based courses. She is moderator for BTEC and Equal Opportunities Officer for the London branch of the Institute of Housing.

Veronica Coatham is currently Head of the Housing School at the University of Central England in Birmingham. She began her career in housing on graduating from University College Swansea and has had experience of working in local authority housing departments and also at the Institute of Housing where she was employed as an education officer in the mid-1980s. She is a fellow of the Institute of Housing and a BTEC moderator.

Jane Darke, Ph.D. studied architecture and sociology in the 1960s and has been teaching, researching and writing about housing for many years. She was a housing manager for four years and now teaches housing in the School of Urban and Regional Studies, Sheffield Hallam University. She worked with other women collectively as the Matrix Book Group to write *Making Space: Women and the Man Made Environment* (1984).

Perminder Dhillon-Kashyap has a B.Sc. in psychology and an M.A. in film and video. She has worked with Black women and young people in the field of media – researching television documentaries as a journalist and freelance trainer.

Anne Douglas is currently employed as a trainee solicitor. Prior to this she worked in housing, in the voluntary sector, for Tyneside

Housing Aid and Housing Aid for Youth advising young people on their housing rights.

Karen A. Franck is Associate Professor at the New Jersey Institute of Technology in Newark, New Jersey, where she holds a joint appointment in the School of Architecture and the Department of Social Science and Policy Studies. She has a Ph.D. in Environmental Psychology from the City University of New York and is past chair of the Environment Design Research Association. With Sherry Ahrentzen, Karen edited *New Households, New Housing* (Van Nostrand Reinhold, 1989).

Rose Gilroy, B.A., M.A., MCIH was educated at the Universities of Nottingham and Exeter before working in local government as a housing officer. She moved in 1988 to the University of Newcastle where she teaches housing policy and issues of equal opportunities. She has published on neighbourhood participation in urban regeneration, the housing needs of young people and equal opportunities for planners.

Janet Hale has worked in the local authority, housing association and higher education sectors since graduating in psychology in 1981. She currently works in the tenant participation and consultancy department of CDS (Liverpool) undertaking training, resource management and research projects. A fellow of the Institute of Housing, she is an external examiner for postgraduate housing courses and is currently researching an M.A., looking at the career cycles of women directors.

Marianne Hood began work with tenants' groups in 1974 and has been secretary of federations in Brent, Tameside and Manchester. A qualified teacher, she was appointed director of the Tenant Participation Advisory Service for England in 1988. She represented TPAS on the reconvened Duke of Edinburgh's inquiry into British housing.

Julia Smailes is a housing officer currently working in Cardiff. She has recently completed the postgraduate diploma in housing administration at Sheffield Hallam University, where she took the housing experiences of lesbians as her research topic.

Roger Sykes is Research and Information Manager with Anchor Housing Trust. After completing a geography degree at the University of Reading in 1985, Roger worked for Oxfordshire Health Authority on a DHSS research project looking at equipment for disabled people. Roger joined Anchor in 1988.

Roberta Woods, BSc., D.Phil. was educated at the University of Ulster. She has worked in local government and for a number of years taught housing management/policy at the University of Ulster. She currently teaches social policy at the University of Newcastle upon Tyne. Roberta has written on various social security issues, housing, and consumer views of health services. She is at present undertaking a major research project on women's access to local authority housing.

Tom Woolley qualified as an architect at the University of Edinburgh in 1970/71. With a Ph.D. from Oxford Polytechnic, he is currently Professor of Architecture at Queen's University Belfast. He is a founder member of the Association of Community Technical Aid Centres and the Ecological Design Association, a member of the Board of Belfast Improved Houses and the Council of the Royal Society of Ulster Architects, and an external examiner for the part-time housing course at Stirling and Heriot-Watt Universities.

Chapter 1

Introduction

Roberta Woods

The idea for this book arose out of a conference held at the University of Newcastle upon Tyne in 1991 to explore current issues for women in the delivery and management of housing services. It was the view of the conference organisers that the conference would explore an agenda that was relevant to the 1990s. It was also the intention that the papers delivered at the conference would present, as far as possible, a holistic view of women's housing concerns. The conference was divided into three parts to consider women as consumers, providers and protesters of housing services. This gave participants an opportunity to consider how women access a range of housing tenures; innovative housing designs; the ways in which various groups of women have different experiences of housing; how appropriate housing education is for women; career prospects within housing management, and the role that women play in campaigning for better housing.

This book draws on the themes that were identified at the conference and thus includes chapters which address the wide-ranging nature of women's housing issues. There is, however, another reason for this broad-based approach. Both the editors are actively involved in housing education and are thus acutely aware of the need to produce a textbook which will seek to introduce students to several key issues in housing for women.

Since 1980 a number of publications have sought to address the issue of gender inequality in the housing market. This seems to have become a particularly crucial issue during the 1980s, given the thrust of government policy towards support for owner occupation with a residual 'social' or 'independent' sector to accommodate those not able to purchase on the open market.

Literature on this subject to date has considered issues which disadvantage women in gaining access to owner occupation (Shelter Briefing, 1987; Brion, 1987). But it has gone further to outline aspects of policy which undermine women's needs or, at a more theoretical level, has sought to examine the role of housing form in reproducing stereotyped gender roles and the dominance of the nuclear family (see Watson, 1986, 1988; for a different perspective see Saunders and Williams, 1988; and Saunders, 1989).

Within the literature much attention has been given to the structural disadvantages that women face and which place them in a weak position with regard to the provision of housing. Empirical evidence on women's actual experience of housing and how this relates to housing management practice is rare. The literature focuses on the barriers to owner occupation and difficulties faced in gaining access to council housing.

Income differentials between the sexes are seen as particularly important in limiting access to owner occupation (Morris and Winn, 1990; Brion, 1987). Morris and Winn also suggest that it is the lack of a male breadwinner which appears to be the crucial factor in restricting access. Policies of lending agencies which might discriminate directly against women have also been highlighted (Watson, 1988; Equal Opportunities Commission, 1978). However, direct discrimination by, for example, treating women's earnings differently to those of men, or not treating women seriously when they apply for a mortgage, has been challenged by Glithero (1986). Improvements in the situation with regard to mortgage lending have been noted by the Nationwide Anglia Building Society (1989).

In respect of council housing, attention has focused on housing allocations policies which might discriminate against women (Kelly, 1986; Brailey, 1985); for example, a lack of sensitivity about offering dwellings in close proximity to previous violent partners, or ignoring the needs of single women without dependent children. The particular problems faced by women following relationship breakdown have been documented by Logan (1986) and the Institute of Housing and the Scottish Homeless Group. Problems can vary from local authorities requiring proof of separation before rehousing to the type and unsuitability of the housing offered. This may on a temporary basis include being placed in bed and breakfast accommodation which Conway (1988) has shown to be particularly unsuitable for families and a cause of ill health.

Given the difficulties which women may face with regard to housing on the breakdown of a relationship, a number of women become homeless or apply to local authorities to be rehoused under the homeless persons legislation (Brailey, 1985; Watson and Austerberry, 1986; Logan, 1986; Morris and Winn, 1990).

Much of the women and housing literature broadens into the debate within urban geography to consider the relationship between gender roles and the form of the built environment. Of importance here are planning policies, architecture and design (Bowlby et al., 1982; Little et al., 1988). These studies reflect upon safety and design and architectural impediments to supportive living such as nuclear family housing or deck access high rise. Danger for women is linked to underground walkways and car parks, poor lighting and excessive shrubbery. These point to more theoretical questions relating to gender divisions of space (McDowell, 1983) or the ways in which the home, location and environment curtail or enhance women's choice (MacKenzie and Rose, 1983).

It is also the case that literature has addressed the differing housing needs or different experiences of housing services of young women (National Council for One Parent Families, 1989), elderly women (Darke, 1987) and Black women (Rao, 1990).

Sexty (1990) examines the economic and social constraints which place women in a disadvantaged position in the labour market and examines the effects of privatisation in housing. Data from the General Household Survey are presented showing that single, widowed and divorced or separated women are more likely to be council tenants and less likely to be home owners than their male counterparts. Similar under-representation in owner occupation among single females was noted by Munro and Smith (1989) using data from the National Child Development Study. Such information suggests that council housing is particularly important for women. How this housing is allocated to them is, then, a significant issue. More recently Muir and Ross (1993) have examined the particular problems that women face in accessing housing in London.

The literature, while addressing a wide range of issues with regard to women and housing, has focused on economic constraints, weaknesses in the law relating to family breakdown, inadequacies in homeless persons legislation, general problems of

allocation systems and latterly to the concentration of women
in the public sector.

This book seeks to build upon the current literature briefly
outlined here and to draw together the diverse strands of women's
housing experiences. The first four chapters consider the general
themes of access, participation and the meaning of home.
Chapters 5–8 address the housing concerns of specific groups of
women. Chapters 9 and 10 look at education, training and employment issues. Chapters 11 and 12 present innovative housing
designs. General conclusions are contained in Chapter 13.

Jane Darke, in Chapter 2, considers the meaning of home.
Criticising Saunders' assertion (1990) that the meaning of home as
a nurturing and safe haven is the same for men and women, she
argues that women value their homes in a particular, contradictory, way. Women contribute much greater time and energy to
housework than men. Although most women experience this domestic role as burdensome, it is difficult to relinquish because of
women's strong identification with the home: 'the sense of exploitation can coexist with pride at performing at least some aspects of
the role well' (this volume, p. 23). This mixture of love and
resentment, Darke suggests, is common to most women. There is
no false consciousness in women feeling pride and pleasure in
housework, and indeed many women are denied this by financial
pressure to go out to work. In the same way, male violence and
bad housing mean that many women are denied the experience of
home that others enjoy.

Rose Gilroy (Chapter 3) considers women's access to housing,
establishing that women find it difficult to become owner occupiers
without a male partner. Most women are unable to win this 'mark
of success' on their own even if they are economically active: they
must find their prince before they can get their palace. Using data
from a study by McCarthy and Simpson (1991), Gilroy observes
that upon divorce among owner occupiers, custodial fathers have
greater staying power in the tenure compared with custodial
mothers, who are more likely to end up renting. This finding is
linked to women's poorer wages and limited access to full-time
waged work given caring and domestic responsibilities. She also
suggests that women caring for dependants in owner occupation
are less likely to inherit property in the future because of pressure
to release equity to pay for formal care. Problems of debt too –
both mortgage and rent arrears – may be affecting women dispro-

portionately. This problem is made worse by many women having to buy or rent housing with greater maintenance and running costs, further compounded by women's lack of DIY skills. Gilroy concludes that, if women are to be helped into owner occupation, more funding is needed for equity share and mortgage rescue schemes. However, women will always need the option of responsibly managed, quality and affordable rented property.

Women and tenant participation is the subject of the chapter by Hood and Woods (Chapter 4). Although women are not well represented among housing decision-makers, they are often very active in tenants' associations. The council tenant's right to be consulted was made law by the 1980 Housing Act. By the late 1980s over two-fifths of local authorities had formal methods for involving tenants in decisions, while four-fifths had informal methods. In giving tenants a right to opt for an alternative landlord (a right, incidentally, that few tenants wanted), the 1988 Housing Act was a strong stimulus for local authorities to develop tenant participation within a 'customer care' framework, encouraged further by the Citizen's Charter. However, in reality tenants' influence over decisions remains weak. Tenants, for example, have no rights to be consulted about rents or other financial matters. Hood and Woods suggest that this will remain the case until tenants can make a collective input to decisions, with their representatives having negotiation rights. This degree of involvement has to be supported by landlords, with tenants' confidence being built up within a culture of participation. It is often women who get together first, but they face particular problems. Among these are childcare and dealing with intimidating and bureaucratic procedures. As a result, men often end up leading the groups.

Roger Sykes (Chapter 5) considers the position of older women, a large and growing part of the population, many of whom are living alone. It is estimated that over 1.3 million older people are living in poor housing conditions in urgent need of assistance. Most of these people are women. Despite recent promotion of the 'Woopie' (well-off older person) concept, only about 20 per cent of the older population could be considered to be comfortably off; 40 per cent are poor and extremely disadvantaged. Older women are much less likely to have occupational pensions; those who have receive markedly less, and older women are likely to have lower savings. Older women are more likely to suffer disabilities than older men, as well as loneliness and social isolation. Roger Sykes reminds us that around 90 per cent of older people live in general housing, with only 10 per cent in sheltered housing, residential or nursing homes. However, old age, disability and poverty often make desirable a move out of general housing into manageable, affordable accommodation, possibly with care. Reviewing older people's housing options, he describes in detail the range of provision developed by the housing association Anchor, England's largest provider of housing with care. Affordable sheltered housing, locally available, together with an option to access 'Staying Put' or 'Care and Repair' services, if these are what is needed to stay in general housing, are particularly important components of a housing policy for older people.

Perminder Dhillon-Kashyap considers the housing needs of Black women in Chapter 6. Black women are as diverse a group as any other but have in common experiences of racism and sexism. Discrimination in owner occupation, private renting and council renting is a long-established and widespread experience. This discrimination is associated with racist propaganda and racist attacks, as evidenced in the September 1993 local election victory for the British National Party on the Isle of Dogs. Women are particularly affected by concerns about their personal safety on the streets as well as in their homes. Escaping domestic violence can mean deportation if this entails breaking the '12-month rule'. Checks on the immigration status of Black people by housing authorities mean that many are being required to *prove* their entitlement to services. These issues add to the housing problems caused by higher unemployment among Black people. Homelessness or domestic violence is often more difficult for Black women, who face discrimination, a lack of awareness about their rights and a lack of support. Most metropolitan housing authorities fail to maintain monitoring systems, despite continuing evidence of discrimination in accessing council housing.

Douglas and Gilroy (Chapter 7) show how the emphasis of housing and social security policy on the nuclear family has disadvantaged young people, making them vulnerable to homelessness caused by poverty and relationship problems which, in the case of young women especially, is largely hidden. Few young people who are not parents receive priority status under Britain's homeless legislation; it is also difficult for many young people to find work, social security rules discriminate against them, and a combination of low wages and benefit tapers traps many in poverty. In theory, the 1989 Children Act should improve this situation, but the authors found no evidence of this from their work in the north east of England. Homelessness among young women is less public than among men, typically arises from relationship problems and is 'solved' through private channels. Pregnant young women under 18 are particularly vulnerable because they are denied Income Support until within 11 weeks of having the baby. Douglas and Gilroy discuss some housing initiatives in Newcastle upon Tyne designed to respond to these problems, including the 'First Move' scheme which provides supported furnished accommodation for young people.

Smailes (Chapter 8) returns to a critique of the dominance of 'families' in housing policy, this time showing how lesbians' needs are either ignored or suppressed. Lesbians face discrimination and harassment, leading to particular problems of rejection by relations, and problems in sheltered housing, residential homes and hostels. Lesbian relationships are not recognised in law, denying succession rights, for example, and fears about lack of confidentiality mean that their needs remain invisible. A recent National Federation of Housing Associations survey revealed that, while two-fifths of housing associations targeted lesbians and gay men in their policies, few actually monitor lettings and job applications. Smailes found from discussion with lesbian women that their choice is for women-only housing and to live in areas with other lesbians.

Equal opportunities in education and training for housing is discussed in Brion's chapter (9). Many women are employed in housing services but few are to be found among the senior posts. Although there is evidence that this is changing, it may actually reflect achievements during the 1980s that have not continued in the 1990s. Brion argues that men's dominance in senior housing posts reflects male power in education and training, particularly in building-related areas, reinforcing the idea that leadership, management and technical subjects belong to men. Equal opportunities policies often fail to address the aspects of behaviour and curriculum content which perpetuate the subordination of women. Monitoring is often weak and performance indicators are often not used. Brion sets out a checklist which can be used by employers seeking training and which promotes equal opportunities. This includes such checks as whether the training institution provides guidance and training to staff about non-discriminatory practice, and whether men and women are encouraged to contribute equally to class discussion, including sensitive discussion of issues of unequal participation. Within the Institute of Housing (IOH), the equal opportunities working group has brought about significant changes with, for example, the Institute committed to 50 per cent of its members of council being women by the year 2000. The part-time route to IOH qualifications has been retained on equal opportunity grounds.

In 1992 there were only 16 women chief housing officers in England and Wales. This, however, was six more than Coatham and Hale found in 1989 when they interviewed women in these positions. In their chapter (10), they point out that the critical career years for achieving management responsibilities are when women experience their most demanding family commitments. The literature suggests that among the factors associated with women getting to the top are having a mentor, being childless or divorced, minimising career breaks, an urban base and hard work. Coatham and Hale found that women chief housing officers had also made definite decisions to achieve their positions, and had made remarkable progress on the way. The support of both parents appeared to be influential, as appeared to be being the eldest child 'exercising power in the home'. Organised domestic arrangements seem to be critical, although 'there are few women prepared to trail blaze to the top with children'. Long working days and having to 'programme in' some leisure time were common.

Franck (Chapter 11) discusses the US experience of seeking alternatives to the single family house and neighbourhood, an ideal often enforced in law by zoning ordinances. Neighbourhoods of 'happy families' are the American dream; the home as a place of hard work, conflict or financial worries is denied. This ideal relies on women working in the home and transporting children and other dependants to where they need to be. Franck reviews alternative residential developments in the States. These, for example, combine common spaces with private spaces, space to live, work and possibly rent out, and private rooms with adjacent collective day care. She identifies the need that many households have today for housing with integrated support services which is transitional rather than permanent. There is also growing interest in America in 'cohousing'. This provides individual dwellings but combines them with common spaces and shared facilities, encouraging interaction between households and the sharing of daily chores. Franck concludes, however, by noting the prejudice which exists among regulatory authorities and within neighbourhoods against these kinds of housing arrangements. Woolley continues discussion about suitable housing in his chapter (12).

For him, the battle against past housing failures and a male-dominated construction industry continues to be fought by tenants' associations and community groups, frequently led by women. He warns against tokenistic user participation in design and development, citing instead examples of full involvement giving rise to imaginative ideas for communal facilities, flexible plan forms, covered communal spaces, clusters of housing units, self-building and 'green design'. Housing design can support alternative ways of living not based on stereotypical concepts of the family and the role of women.

Finally, Gilroy in Chapter 13 draws attention to the fact that women's housing needs are still too often neglected. This is particularly important with regard to two key areas of policy development – affordability and quality. These issues need to be addressed from a gender perspective or vital elements in their measurement for women will be lost. There is, of course, a vital need for more research to be carried out into the housing needs of women, and we still await the widespread availability of housing design solutions that move beyond conventional family structures.

It is hoped that the production of a book of this nature contributes to a literature which seeks to ensure that women's issues in housing are not obscured in the more general debates about housing provision, production and management.

REFERENCES

Bowlby, S., Foord, J. and MacKenzie, S. (1982) 'Feminism and Geography', *Area*, Vol. 14: 19–25.
Brailey, M. (1985) *Women's Access to Council Housing*, Occasional Paper No. 25, Glasgow: The Planning Exchange.
Brion, M. (1987) 'The housing problems women face', *Housing Review*, Vol. 36, No. 4: 139–40.
Conway, J. (ed.) (1988) *Prescription for Poor Health: The Crisis for Homeless Families*, London: SHAC.
Darke, J. (1987) 'Report from housing workshop', *Planning and Housing Policies: Their Effect on Women*, Sheffield: Sheffield Centre for Environmental Research.
Equal Opportunities Commission (1978) *It's Not Your Business, It's How*

the Society Works: The Experience of Married Applicants for Joint Mortgages, Manchester: EOC.

Glithero, A. (1986) 'Lending to women – dispelling the myths', *Housing Review*, Vol. 35, No. 6: 202–3.

Kelly, E. (1986) 'What makes women feel safe?', *Housing Review*, Vol. 35, No. 6: 198–200.

Little, J., Peake, L. and Richardson, P. (1988) *Women in Cities*, London: Macmillan.

Logan, F. (1986) *Homelessness and Relationship Breakdown: How the Law and Housing Policy Affects Women*, London: One Parent Families.

McCarthy, P. and Simpson, R. (1991) *Issues in Post Divorce Housing*, Aldershot: Gower.

McDowell, L. (1983) 'Towards an understanding of the gender division of urban space', *Environment and Planning D: Society and Space*, Vol. 1: 59–72.

MacKenzie, S. and Rose, D. (1983) 'Industrial change, the domestic economy and home life', in J. Anderson, S. Duncan and R. Hudson (eds) *Industrial Spaces in Cities and Regions*, London: Academic Press, pp. 155–200.

Morris, J. and Winn, M. (1990) *Housing and Social Inequality*, London: Hilary Shipman.

Muir, J. and Ross, M. (1993) *Housing the Poorer Sex*, London: London Housing Unit.

Munro, M. and Smith, S. J. (1989) 'Gender and housing: broadening the research debate', *Housing Studies*, Vol. 4, No. 1: 81–93.

National Council for One Parent Families (1989) *Young Single Mothers Today: A Qualitative Study of Housing and Support Needs*, London: NCOPF.

Nationwide Anglia Building Society (1989) *Lending to Women 1980–1988*.

Rao, N. (1990) *Black Women in Public Housing*, London: Black Women in Housing Group.

Saunders, P. (1989) 'The meaning of home in contemporary English culture', *Housing Studies*, Vol. 4, No. 3: 177–92.

Saunders, P. (1990) *A Nation of Home Owners*, London: Unwin Hyman.

Saunders, P. and Williams, P. (1988) 'The constitution of the home: towards a research agenda', *Housing Studies*, Vol. 3, No. 2: 81–93.

Sexty, C. (1990) *Women Losing Out: Access to Housing in Britain Today*, London: Shelter.

Shelter Briefing (1987) *The Impact on Women of National Housing Policy since 1979 and Prospects for the Future*, London: Shelter.

Watson, S. (1986) 'Housing the family: the marginalisation of non-family households in Britain', *International Journal of Urban and Regional Research*, Vol. 10, No. 2: 8–28.

Watson, S. (1988) *Accommodating Inequality: Gender and Housing*, Sydney: Allen & Unwin

Watson, S. and Austerberry, H. (1986) *Housing and Homelessness: A Feminist Perspective*, London: Routledge & Kegan Paul.

Chapter 2

Women and the meaning of home

Jane Darke

The core argument of this chapter is that there is a distinctive relationship between women and their homes; that women value their homes in a particular way. Our feelings are a mixture of affection, reciprocated towards the home as a nurturing environment, and resentment at the demands of the home. I have felt this intuitively for some time (Darke, 1989) but to substantiate these ideas is a long project that does not lend itself to conventional research methods. In attempting to muster evidence to support my contention, I have drawn on a variety of sources: writings by others about women and men in their homes, writings by academic feminists, poetry, and interrogation of my consciousness and that of friends about our feelings about home and our roles within in.

The methodology of this exploration is interpretative rather than positivistic, and makes use of subjective material. In some senses it complies with the model of feminist research put forward by Liz Stanley and Sue Wise (1983). They argue that feminist research should begin with the personal, should name women's own experience, recognising that women 'are social beings . . . [who] interact with other people at all times, either physically or in our minds' (p. 166). However, unlike Stanley and Wise's model, it does not reject research carried out on women 'subjects' by a researcher. Stanley and Wise believe that such accounts are based on 'fictitious sympathy' if the researcher claims to share emotions but has not shared experiences. This exploration does use other people's accounts of their feelings about the home and domestic labour, and accounts collected by other researchers, but recognises that the user of such sources must consider critically how such material is collected, the power relationship between research(er) and subject, and thus the status of the knowledge presented.

Within the overall theme of the meaning the home holds for women, there are many sub-themes. Some touch on issues which other academic writers have discussed, and some previous writers will be challenged. Peter Saunders (1990: 304–10) has argued that men and women view the home in essentially the same way: as a haven for relaxation and loving relationships (and, furthermore, that these benefits are only fully available to home owners). He also makes some highly contentious points about the division of domestic labour and the skill levels involved. On the other hand, some feminist writers have seen the home only as a site of oppression, exploitation and male domination (Comer, 1974; Barrett, 1980). Such interpretations derive more from marxist theory than the expressed views of the women whose experience is being theorised (Kaluzynska, 1980). In assigning wholly negative meanings to women's work in the home, are marxist feminists undermining a role which is for many women a strong source of identity, pride and satisfaction? It is also important to look at the way gender relations and expectations of women and men in the home are changing.

This chapter is also an implicit criticism of the narrowness of much writing on 'women and housing', which has concentrated on women's disadvantaged access to housing. Any competent writer or student of housing and social policy can collect data on tenure by gender of head of household, average earnings of men and women, evidence on policies and practices by local authority housing departments and other agencies, and can then construct a well-supported argument on women's poor access to housing. This has been competently done quite recently by various writers (e.g. GLC, 1986; Resource Information Service, 1988; Watson, 1988; Sexty, 1990; Morris and Winn, 1990: chapter 4; Clapham et al., 1990: 71–8). There are few texts on 'women and housing' that look beyond issues of material disadvantage (but see Matrix, 1984; Madigan et al., 1990).

This chapter looks next at the writings of Saunders (1990) on gender and the meaning of home, and at gender differences in domestic labour, and then at other interpretations of women's work in the home and their feelings about it.

Saunders (1990: 304–10) has asserted that his data show no gender differences in the meaning of home. The question he asked was a single, rather bald one: 'People often distinguish between "house" and "home". What does the home mean to you?', at the end of a long and at times loaded interview schedule. Among the 'subjects' of his survey two sets of meanings predominated, with 35 per cent of men and 37 per cent of women mentioning family, love and children, and 23 per cent of men and 27 per cent of women mentioning comfort and relaxation.

It is noticeable that on both these issues more women mentioned them than men: this is true of most of the categories of meaning that Saunders lists. Altogether, the women who responded to this question produced an average of 1.4 'meanings' each while the men averaged 1.1. This suggests at the very least that women's views of the meaning of home are more complex than men's. This is borne out by Gurney's work, discussed below.

In fact, it is clear that not much reliance can be placed on Saunders' findings. It is a fairly elementary methodological point that a single question in a large-scale survey is *not* the appropriate way of exploring complex and subtle areas of meaning. The reader is not even told about the circumstances of the interview. For example, were couples interviewed together? If so, it is not likely that the women would feel comfortable about voicing an ambivalent set of meanings including the burdensome nature of domestic labour. One is reminded of the explanation for the under-reporting of domestic violence in the British Crime Survey as being owing to 'the assailant being in the same room at the time of the interview' (Hough and Mayhew, 1983, quoted in Morris and Winn, 1990: 132). We are told that the interviewers in Saunders' study included both men and women but there is no discussion about whether the results differed according to gender of interviewer.

Gurney (1990; 1991) adopted a methodology more suited to exploring subtle nuances of meaning. He employed a multi-staged approach where a sub-sample was taken from an initial large sample, for semi-structured and unstructured interviews around issues of meaning. These reveal the intricacy of feelings which one could sense were there waiting to be discovered given an appropriate methodology. The home was certainly seen as part of the emotional sphere and, particularly for women, was inseparable from significant life-events that had taken place there, from childbirth to relationship breakdown. Gurney, interviewing in 1990, discovered another strand not reported by Saunders' interviewers in 1986: a set of negative feelings associated with the mortgage being a burden and the home being a 'millstone'. These attitudes towards an owner-occupied home were also seen in some of Cherrie Stubbs' Sunderland sample (1988: 152). Gurney found a further set of meanings associated with pride and a sense of security in attaining a home of one's own, which appeared to be particularly significant for single women householders.

WHO DOES HOUSEWORK?

It has often been presumed that women experience the home differently from men because they are seen as responsible for housework. As the Matrix authors say: 'even when others contribute to this work, the primary responsibility remains with women. We are conscious of its demands at all times' (1984: 2). There is a series of issues to be discussed here: who does the domestic labour; what level of skill does it entail; is male participation on the increase; do women experience domestic labour as burdensome; whether or not they find it burdensome are they, objectively speaking, exploited or oppressed and, if so, by whom; does this apply to all women or only those living with a husband and/or children; how does this affect our feelings about our homes?

There have been various surveys of the division of labour in the home on which it is not necessary to report in detail (see, for example, Pahl, 1984; Gershuny, 1983). Saunders (1990: 304) discussed some journalistic surveys of 'who does what' and concludes that male participation in housework is still low, pointing out that this was also the case in his survey. However, he appears not to grasp the extent of the inequity when he states that 'work is *divided* . . . women do not do everything' (emphasis in original text). Saunders' data cover nine tasks, and figures are given separately for three occupational groups. In only one of these 27 instances (car maintenance in social groups I and II) is the man mainly responsible in more than 50 per cent of cases. There are 11 cases where women are mainly responsible and 15 where the 'median position' is joint responsibility. See Table 2.1.

Saunders and Williams (1988) elsewhere attempt to theorise the patterns that emerged by suggesting an inside/outside division (men tending to do work to car, garden and exterior home maintenance). This has a certain plausibility but reinforces the inequality in that most 'outside' tasks are occasional rather than daily.

Table 2.1   Who does the work in the home? (class groups combined)

Main responsibility %	Women	Both	Men
Cooking	79	12	9
Washing up (dishes)	52	33	14
Shopping	53	36	11
Washing and ironing	84	10	6
Cleaning the house	68	26	6
Gardening	26	27	48
Car maintenance			
(car owners only)	17	10	73
Window cleaning	43	18	40

Source: Adapted from Saunders (1990: 305)

They also suggest that 'symbolically polluting' tasks fall to women. Again this is plausible, although it might be hypothesised that a certain class of rarely occurring polluting tasks tend to be carried out by men: removing dead mice and unblocking outside drains, perhaps (with considerable variation in 'who does what' here, I suspect).

Saunders and Williams suggest that tasks low on information and high on energy tend to be mainly done by women, and vice versa for male tasks. Their own examples relate to preparing Sunday dinner as against carving the joint, and mending tears in clothing as against electrical fuses. The issue of which of these tasks uses more energy is not particularly clear cut, but the contention that the 'male' tasks imply greater knowledge/ability demonstrates only how little Saunders and Williams understand the skill levels involved. I will not detail the complexities of preparing a meal such that the right quantities of appropriate foods reach the table cooked to the right degree at a particular time, but I will review the skills involved in mending clothing.

It is necessary to judge whether to patch or join up; if patching, the colour and strength of fabric to use in relation to the garment fabric; thread colour and strength; how to avoid fraying of patch and tear; threading a needle; what stitch type is least visually intrusive, and so on. (I am slightly anxious that revealing that I know so much about a task defined as low in knowledge by male professors will damage my credibility as an academic. However, as a feminist I am affronted that the highly developed and complex skills involved in virtually any domestic task one could name are not perceived as such: this ignorance can only undermine the validity of theorising based on such false premisses. I am also *proud* that I have a basic level of competence at such tasks, and will return to this issue later.)

Of course, the undervaluing of women's skills is a familiar phenomenon in paid work. There have been many industrial disputes centring on skill levels and pay differentials. Despite some victories it is still often the case that 'the sex of those who do the work, rather than its content . . . leads to its identification as skilled or unskilled' (Phillips and Taylor, 1980). The same appears to be true of domestic labour.

Domestic labour, then, falls overwhelmingly to women; a high level of skill is involved in the tasks; *if* male participation is increasing it is at a very slow rate and from a low baseline. Furthermore, any increase in men's participation in stereotypically female tasks such as cooking and ironing may well be offset by more women taking on 'male' tasks such as 'do-it-yourself', painting and decorating, and car maintenance.

FEELINGS ABOUT HOUSEWORK

There was considerable debate in the 1970s on 'domestic labour': was it productive work; did it produce surplus value; who exploited the labour of the housewife; should she be paid; did women as housewives constitute a class? As the vocabulary suggests, the debate was largely among marxists, challenged by the rise of the women's movement into discovering revolutionary potential in women's domestic role. The various positions held have been usefully summarised by Kaluzynska (1980) but by the 1980s the issue had virtually left the agenda, to be replaced by issues such as women's access to employment and career opportunities, and problems of male violence.

This chapter does not seek to reopen the debate although Pugh (1990) has recently attempted to do so. I have much sympathy with Stanley and Wise (1983) when they say that 'whether domestic labour is productive or non-productive . . . seems relevant only . . . to rather few marxist feminists. Marxist feminism's "feminist theory" is *only* "women are oppressed"' (pp. 43–4, emphasis in original).

However, to understand women's own experiences of home and their work within it, it is important to disentangle different empirical questions. We have established that women contribute much greater time and energy to housework than men. Do they experience this labour as oppressive? If so, and given that *most* paid jobs include some tedious, frustrating or tiring elements, does that mean women are dissatisfied with the role of housewife and homemaker? How do they decide what standard of work to aim at and what standard is tolerable? What are the satisfactions of the housewife role? This chapter examines at length what women have said about their work in the home, how researchers have interpreted this, and speculates on the psychological ties between women and the home, before returning to a further brief discussion of marxist positions on domestic labour.

Ann Oakley (1974) talked at length to 40 women aged 20 to 30 with young children about their work as housewives (one was also in full-time employment and five did paid part-time work). Answers to the direct question 'do you like housework?' showed that half claimed to dislike it, a quarter to like it while a quarter 'didn't mind'. Oakley's own interpretation, based on the way the women she interviewed described their work, was that 30 per cent were satisfied and 70 per cent dissatisfied. Two-thirds were assessed as being satisfied with their childcare task. These bald statistics conceal some sensitive interpretations (see original text) including the social pressures for acceptance of the housewife and carer role: pressures which are now perhaps more equivocal.

It was clear that the women in Oakley's sample set their own standards of cleanliness to be attained, sometimes in accordance with or in reaction to their mother's standards. In no case did the husband attempt to set a standard, and husbands rarely praised their wives' work. Oakley shows that any 'job satisfaction' arises not from the attitude of the husband as 'boss' but from self-reward at maintaining standards which, although set by the woman herself, are then assigned an objective reality. The pleasure felt by some women seems real enough: 'I just like [cleaning]. I'm happy when I'm doing it . . . I love doing my own washing. I don't believe in washing machines' (Oakley, 1974: 109). The hatred of the housewife role voiced by other writers is equally vivid:

> Every day is exactly like the previous one with nothing to look forward to and nothing to look back on, with the certain knowl-

edge that she is standing still while the rest of the world moves on. She lives a half life which is marked out by pseudo events.

(Comer, 1974: 99)

More recently, however, there has been more subtle analysis of the woman's role in establishing the social standing of the household and its members. This is not 'status striving', rather the competent maintenance of normative expectations. Many of the contributors to Allan and Crow's (1989) collection of essays entitled *Home and Family* describe this process.

Marion Roberts interviewed women who had moved into a small development of council flats built in the early 1950s. It was through the high housework standards of these women, along with the resident caretaker, that the respectability and status of the estate were maintained:

the 'success' of the estate was hard-earned, by the women and the men. The women had to combine housework, waged work and childcare, all the while attempting to suppress any visual signs of domestic work such as washing, noisy children playing, and clutter and mess. By inference, the men had to work long hours [to afford higher rents].

(Roberts, 1989: 46)

Roberts also shows that each woman has her own standards: for example, one woman considered it acceptable to dry washing on the balcony despite other women seeing this as evidence of lower standards, but would not put out washing on a Sunday.

Pauline Hunt used the imaginative research method of lending her subjects polaroid cameras to take pictures of their homes, which then formed the agenda for interviews. The homeworker:

wants her home to be seen (publicly scrutinised) as clean and tidy, and at the same time she wants it to be experienced (privately appreciated) as free and easy. The houseworker's practice tries to reconcile these contradictory objectives. When family members are home she does not veto their disorder-creating behaviour but she constantly re-establishes order since . . . 'You never know who may come'. If someone should call and see the disorder, she has no doubt which family member will be held to have fallen short of her duty . . .

[T]he wider community makes itself felt . . . not only through the pressure of expectations exerted upon her by neighbours, her husband's kin and work associates, and the parents of her children's school friends and their adult associates. There is also a more general pressure resulting from her perception of her domestic role in society, which tends to be reinforced through cultural images and messages related to domestic practice.

(Hunt, 1989: 69, 68)

Yet this role is valued, and its partial loss is regretted when family circumstances change:

Mrs Poskier's home making skills were not without an audience, for her teenage children were still living at home at the time of the interviews. When they leave to establish homes of their own, her domestic management skills may well be experienced as less satisfying and more meaningless. Then, like Mrs Scott, most of her satisfaction may come from visiting her children in their own homes. . . . Since Mrs Scott is not the focal point of her children's households, and her own house lacks an appreciative audience, she comes close to being emotionally homeless.

(Roberts, 1989: 73)

However, the return of non-resident children may also be a source of tension, as a couple interviewed by Jennifer Mason demonstrates:

Pete . . . when the kids come home, they're not untidy or anything like that but, naturally, they shift here, they shift the cushions, . . . they've only been here five minutes and it looks as though a bomb's hit it. . . . And when they go back you think 'phew!' Evie: Yes, you think 'ooh it's lovely and tidy again'.

(Mason, 1989: 107)

Mason interviewed couples in their 50s and 60s, married for several years. She talks about little discussed aspects of the woman's role in the family home: 'worrying about, noticing, monitoring, feeling guilty, promoting and ensuring everyday well-being and so on, are aspects of a particular kind of labour which sits more neatly under the heading "caring about", than "caring for". It is mental labour' (Mason, 1989: 119).

There is also the public dimension described by Roberts and Hunt:

the production of the home for public scrutiny . . . does not mean simply making the physical environment look nice for visitors, although that is an element of the procedure. Rather, it means producing the home – including its members – for public approval. In other words, producing the home means upholding in some public sense the status of the family household . . . I am suggesting that the women in my study were engaged continually and routinely in this kind of process.

(Mason, 1989: 120)

Most of the women interviewed by Roberts, Hunt and Mason had no private space of their own, but hardly understood a question from Hunt asking whether they missed this, regarding the whole house as their domain (Hunt, 1989: 71–2). Virginia Woolf's demand for 'a room of one's own' (Woolf, 1929) does not appear to find an echo, although Gurney's (1991) as yet unpublished material includes accounts of how his subjects (of both genders) 'escape', through reading, going into the garden or even visiting the toilet.

Even without personal space, however, *control* over one's time and space is clearly important and was valued by Oakley's subjects. It might suggest that having control over the time scheduling of the use of the home obviates the need for one's own space. The loss of such control, through the presence of visitors or a change in family circumstances, can be stressful. One of the Matrix authors, certainly not the stereotypic house-proud housewife, found that her feelings about the state of the house changed after her first child was born:

[W]hen Kim was just a couple of weeks old Mike came back from work one evening to be greeted by a barrage of fury because the house was a mess, the dishes hadn't been done the night before, the nappies hadn't been washed and the place was in chaos. He was quite surprised. We had always lived in chaos so what was new? I realised then that what was new was that I no longer had complete control of my time . . . I couldn't just put aside two hours to blitz the mess when it got more than I could bear.

(Foo, 1984: 132–3)

MY HOME, MY MOTHER, MYSELF

There is a rich vein of evidence, when we look beyond the oversimplifications of the mass survey, of women's complex relationship to their homes. It is clear that, even when the domestic role is experienced as burdensome, a mesh of internalised social constraints makes it almost impossible to relinquish: the identification with the home is too great. The socialisation process that gives rise to this leaves an imprint at the psychoanalytic level. To lead into a discussion about this I include a poem by Kathleen Jamie (1991) called 'Wee Wifey', which expresses some of these issues more subtly than I possibly could.

I have a demon and her name is
                    WEE WIFEY
I caught her in a demon trap, the household of my skull.
I pinched her by the heel throughout her wily transformations
until
        she confessed
                    her name indeed to be WEE WIFEY
and she was out to do me ill.

So I made great gestures like Jehovah: dividing
land from sea, sea from sky,
                    my own self from WEE WIFEY
(*There*, she says, *that's tidy!*)

Now I watch her like a dolly
keep an eye,
            and mourn her;
for she and I are angry/cry
                    because we love one another dearly.
It's sad to note
            that without
                WEE WIFEY
I shall live long and lonely as a tossing cork.
                    (Reproduced by permission of *New Statesman &*
                        Society, 19 July 1991)

This is where the argument becomes unavoidably speculative, but I would be particularly interested to know whether it resonates with others. I sense that many of us, men and women, identify the home with the fantasy 'good mother', a place where we are accepted for what we are, that nurtures us and restores us so that after a time we can go out into the judgemental world again. Yet along with the home as nurturing mother, there is a whole host of ambivalent feelings, just as most of us have conflicting feelings towards our real mothers. The home may comfort us but it also makes very real demands on us; we resent our dependence on it and resent the imposition of order and routine that we recall from the real maternal home.

It is a sort of *rite de passage* for young adults, especially men, to go 'on the road' and to spend a few months or longer in a situation where they have no regular home or routine. These journeys are celebrated in literature; Kerouac (1958) and Pirsig (1976) are two instances among many. Although these two authors are distinctly post-adolescent, the rebel, free spirit mentality, breaking free of societal and especially parental expectations, is strongly evident. The 'escape' may be from the wife or girlfriend (or womankind in general) but is seen by the escaper as an emphatic rejection of an ordered, settled existence in a *home*.

Our reactions to these self-romanticising travellers may well be mediated by gender and age. Men often seem to envy them, projecting their own fantasies of escaping from a humdrum domesticated existence. I felt this way when younger; now in middle age I tend to see them as rather infuriating delayed adolescents, self-importantly philosophising, obtusely oblivious to the emotional havoc in their wake.

Women cannot so boldly reject the 'home as mother', as we ourselves are, or may become, the mother. This is why we experience at one and the same time irritation at the home 'demanding' that we clean it, and guilt at the fact that we haven't made it clean enough. We feel guilty at refusing to do what would please the mother, at being a 'bad child'. The home is an exacting mother: however much housework we do there is more that 'ought' to be done. For those of us with a relationship to Wee Wifey as ambivalent as the poet's (including, I assume, any woman who has activities she values outside housework) this tension between placing limits on the potentially limitless task of domestic labour and the guilt at so doing is endemic. As I write this chapter I am conscious of a pile of papers and academic ephemera on the floor shouting to be tidied up; I imagine my real mother telling my sisters in shocked tones how I work on a half-made bed in my dressing gown; my real family stick their heads round the door asking if I can get their lunch soon.

Women living without a male partner and/or children may have more space for negotiation with Wee Wifey but I do not believe that she has disappeared with a demon howl. Remember how she and the poet loved each other dearly? The social pressures for maintaining an acceptable standard of home care spring not only from partners and children but from the generalised expectations of society, as Hunt and Mason pointed out in the passages quoted earlier.

This ambivalence also means that the sense of exploitation can coexist with pride at performing at least some aspects of the role well: being able to mend a tear, shop efficiently, make jam, creatively arrange a room or even get the wash whiter than the neighbours (see Attfield, 1989: 233–4). Housework may be seen as 'real work' and equally oppressive, but most women have some areas of competence in domestic tasks which they exercise with pride and satisfaction. Their role in the home is an intense source of meaning.

HOW UNIVERSAL ARE THESE FEELINGS?

Some further doubts may be raised in the minds of some readers. First, do the ideas just discussed purport to describe *all* women? Second, is this not an account of a dying social system, the last gasp of a pre-feminist consciousness which has its roots in women's return to the home after the Second World War?

My own experiences are those of a white, able-bodied, middle-class, heterosexual, married woman, difficult though it is to admit such attributes to a Politically Correct feminist readership. I cannot claim 'fictitious sympathy' (Stanley and Wise, 1983: 166) for women whose experience I have not shared. However, from other people's writings and from personal acquaintance I can make inferences on the extent to which the identification of women with home is common to most women.

There does not appear to be any good evidence that women's feelings about the home and housework differ by social class. Oakley (1974: chapter 4) found no class differences, although she notes social pressures on working-class women to value the house-wife role and on middle-class women to be dismissive. Saunders (1990) discussed gender and tenure but not social class differences in attitudes to the home. My own contacts and interviews with women in their homes, as a friend, as a sociologist and later a housing manager, do not provide any prima-facie evidence of class differences. The authors already cited in *Home and Family* do not describe class differences in taking pride in the home although Pauline Hunt analyses differences in taste in furnishing (1989: 77–8). She suggests that the bright colours, luxurious textures and warmth typical of a working-class home represent 'blanketing as an insurance policy against destitution'. It might be hypothesised that the absence of such amenities owing to *actual* destitution (for example, a household rehoused as homeless but denied a loan for furniture) would be felt more acutely.

Obviously the experience of women will differ according to the ethnic group with which they identify. This is because each group has a distinctive set of cultural expectations towards women and the home, and (for groups other than white British) because racism is gendered. Different cultural groups vary in their expectations of women's economic role, in some cases believing that they should do no paid work at all and in others regarding them as the main breadwinner. However, there are also disturbing similarities in that male domination is evident in all groups, as is the assumption that whatever else a woman may do, she bears the main responsibility for running the home. It is therefore at least plausible that the feelings about the home I have described show cross-cultural similarities.

Lesbians, like heterosexual women, may have many different living arrangements: alone, with a partner, with or without children, in collective households. Again I do not assume that *all* lesbians will share the identification with the home, but one writer on the subject has described why the home is particularly important for her and other lesbians:

> Home represents a great deal to me – touching a woman without a trace of defiance or self-consciousness, feeling 'real' and having my life witnessed. For women who do not have my lesbian networks or public existence as a lesbian, home is the place they can forget the habit of self-concealment and constant vigilance which spring from having a 'secret' self. Home, for all lesbians, is the real or imaginary place where we feel safe, loved and validated.
>
> (Egerton, 1990)

This is clearly a strong challenge to the view put forward by some writers that the notion that the home is a haven for women represents a blinkered fantasy. (In fact, it is difficult to find women writers who have claimed that the home is a haven only for men: the authors cited on this by Saunders (1990), in his vituperative attack on feminist writers on housing, are male, presenting an over-simplified version of rather subtle arguments originally put forward by writers such as Davidoff and her co-authors (Davidoff *et al.*, 1976). However, this does *not* mean that Saunders is correct in asserting that the meaning of home-as-haven is the same for men and women: see below.)

For women who require care by someone else, the possibility of *having* a home may be particularly important because non-disabled writers have been too ready to prescribe forms of collective care and have at times distorted the expressed wishes of disabled people (Morris, 1991–2). Discussing a paper by Dalley (1987) which does just this, Morris says:

> We are invited to treat people's preferences for living in their own homes as a kind of 'false consciousness' arising from what Dalley identifies as the 'familist' ideology which pervades our culture. There is no recognition here that disabled people have often been denied the family relationships that she takes for granted. Insult is then added to injury by the assumption that for a disabled person to aspire to warm, caring human relationships within the setting where most non-disabled look to find such relationships is a form of false consciousness. We are to be denied not only the rights non-disabled people take for granted, but when we demand these rights we are told that we are wrong to do so.
>
> (Morris, 1991–2: 32)

This discussion has shown the possibility, at least, that there are feelings about the home common to most women. The mixture of love and resentment described in the previous section is *not* merely the experience of a particular social segment.

A further objection may point out that, for many women, the home is the site of the exercise of male power through physical violence, rape or mental cruelty (see, for example, Dobash and Dobash, 1980; Hanmer and Saunders, 1984; Binney *et al.*, 1981; Somerville, 1989). In some such cases, physical attack is apparently triggered by a supposed shortcoming of the wife, often trivial such as using the wrong plate or failing to dust the clock. It is clear that the violence is not actually an ill-conceived attempt to change the wife's performance of housework but an arbitrary use of physical strength to intimidate. Any male violence towards women is abhorrent, and particularly so when it occurs in the home. This is precisely *because* it negates the possibility of the home being a secure environment where we have the freedom to relax and be ourselves.

For women the home *is* a haven but in a different sense from the male view of the home-is-haven: it is a haven because the outside world is fraught with threats and other attempts to control us. To be denied the haven of home, to be locked inside with someone who is a threat, is a potent nightmare. This is why it is essential to support women's refuges, and why women's fears of violence must be taken seriously by housing providers. The fact of violence does not contradict the desire for a home as a safe and nurturing environment: it intensifies it.

A WOMAN'S RIGHT TO CHOOSE?

Clearly the negative meanings that the home carries for some writers on housework are not widely shared by other women. It is facile and anti-feminist to dismiss feelings of pride and pleasure in the home as evidence of false consciousness. Neither should we dismiss the feelings of frustration, lack of purpose or captivity in the home voiced by some women. From 1945 until well into the 1960s women had little choice in their lives: the only pattern seen as 'normal' was to marry, have children and become a full-time housewife. A desire for a career, itself seen as aberrant, might be fitted in by postponing a family and/or through a range of stratagems to allow some opportunity to do a job while still performing the normal duties of running a home.

Many determined career women made a conscious decision not to have children; those who combined the two competently were seen as miracle workers (see Rapoport and Rapoport, 1971). My mother returned to teaching in the late 1950s despite having five school-age children, but my father, tired of unironed shirts, was relieved when she had to give up owing to an unplanned pregnancy. Divorce was strongly disapproved of (Wilson, 1980) and even when the husband's behaviour was a problem, the wife's duty (reinforced in every problem page of women's magazines) was to make the marriage work. Thousands of couples stayed miserably together 'for the sake of the children'. Coming from a 'broken home' was even more to be pitied than having a mother who went out to work. No wonder the subjective experience of home for many women was as a prison, and that Betty Friedan's (1963) sigh of 'is this all?' found an echo. Even in the 1950s and 1960s it is clear that the subjective experience of the housewife role was mixed (see Gavron, 1966; Attfield, 1989).

The pleasure and satisfaction that many women obtained from competent performance of the housewife role is now less available because of changed economic circumstances. Ehrenreich (1992) has described these changes perceptively in a retrospective look at the domestic labour debate:

> They (men, capitalists) needed us (women) to do all our tradi- tional 'womanly' things, and . . . would apparently go to great lengths to keep us at it. Well, they don't seem to need us anymore . . . and if this was not completely evident in 1975, it is inescapable today. . . . Mid-1970s feminist theory tended to portray men as enthusiastic claimants of women's services and labour, eagerly enlisting us to provide them with clean laundry, home cooked meals, and heirs. If we have learned anything in the years since then, it is that men have an unexpected ability to survive on fast food and the emotional solace of short-term relationships. . . . Capitalists have figured out that two- paycheque couples buy more than husband-plus-housewife units, and that a society of singles potentially buys more than a society in which households are shared by three or more people.
>
> (Ehrenreich, 1992: 143–4)

Women's participation in the labour market has continued to increase, as has the rate of return to paid work following the birth of a child. A generation ago there was a strong moral pressure on the mother of young children to be a full-time housewife; now there is equally strong economic pressure to remain in a job. The housing system has intensified such pressures, with the rented sector shrinking and therefore becoming harder to get into, and owner occupation now so expensive as to be almost impossible to attain on one income (Bramley, 1991). In theory women have a choice, as they did in the 1950s and 1960s, but in practice the partner of someone in work must work herself, and women with- out partners must work anyway. If the partner is unemployed the poverty trap makes it uneconomic to work, and this is another form of lack of choice.

It is almost as if 'capitalism' noted the cries of marxist feminists that the home was the site of women's oppression, and decided, in a fit of unwonted benevolence, to abolish the role of full-time housewife. But was the analysis of the marxist feminists really serving the interests of most women? This chapter has presented evidence, from and about a range of women, that the home is central to their lives, and that the care of the home and family may be a source of job satisfaction as well as resentment. This is not to deny women's right to jobs and careers, nor to imply that care of a home is some sort of biological destiny, nor to collude with male absenteeism from domestic responsibilities. It is to restate the importance of two things: *choice* for women, and a home that meets her aspirations.

CONCLUSIONS

It is important for those making housing policy or delivering housing services to understand the central significance of the home in women's lives.

Access to a home in which pride can be taken is vital to most women's sense of worth. An unmodernised home which is impossible to keep clean, an insecure home on a high-crime estate, a flat which is accessed by unreliable lifts and fouled and graffitied stairways, a room in a bed-and-breakfast hotel: these are environments which may constitute a 'suitable offer' in strictly legal terms but cannot constitute a home which enhances a woman's sense of herself and of being socially valued. A roof over a head is not enough.

REFERENCES

Allan, G. and Crow, G. (eds) (1989) *Home and Family: Creating the Domestic Sphere*, Basingstoke: Macmillan.

Attfield, J. (1989) 'Inside pram town: a case study of Harlow house interiors 1951–61' in J. Attfield and P. Kirkham (eds) *A View from the Interior: Feminism, Women and Design*, London: The Women's Press.

Barrett, M. (1980) *Women's Oppression Today*, London: Verso.

Binney, V., Harkell, G. and Nixon, J. (1981) *Leaving Violent Men: A Study of Refuges and Housing for Battered Women*, London: Women's Aid Federation England.

Bramley, G. (1991) *Bridging the Affordability Gap in 1990*, Association of District Councils and Housebuilders' Federation.

Clapham, D., Kemp, P. and Smith, S. J. (1990) *Housing and Social Policy*, Basingstoke: Macmillan.
Comer, L. (1974) *Wedlocked Woman*, Leeds: Feminist Books.
Dalley, G. (1987) 'Women's welfare', *New Society*, 28 August.
Darke, J. (1989) 'Problem without a name', *Roof*, March/April: 31.
Davidoff L., L'Esperance, J. and Newby, H. (1976) 'Landscape with figures: home and community in English society', in J. Mitchell and A. Oakley (eds) *The Rights and Wrongs of Women*, Harmondsworth: Penguin Books.
Dobash, R. E. and Dobash, R. (1980) *Violence against Wives: A Case against the Patriarchy*, London: Open Books.
Egerton, J. (1990) 'Out but not down: lesbians' experience of housing', *Feminist Review*, 36, Autumn.
Ehrenreich, B. (1992) 'Life without father: reconsidering socialist feminist theory', in L. McDowell and R. Pringle (eds) *Defining Women: Social Intentions and Gender Divisions*, Cambridge: Polity Press in association with the Open University.
Foo, B. (1984) 'House and home', in Matrix, *Making Space: Women and the Man-made Environment*, London: Pluto Press.
Friedan, B. (1963) *The Feminine Mystique*, London: Victor Gollancz.
Gavron, H. (1966) *The Captive Wife*, London: Routledge & Kegan Paul.
Gershuny, H. (1983) *Social Innovation and the Division of Labour*. Oxford: Oxford University Press.
GLC (1986) *Women and Housing Policy*, GLC Housing Research and Policy Report No. 3.
Gurney, C. (1990) *The Meaning of Home in the Decade of Owner Occupation*, Working Paper 88, School for Advanced Urban Studies, University of Bristol.
Gurney, C. (1991) 'Ontological security, home ownership and the meaning of home: a theoretical and empirical critique'. Paper given at *Beyond 'A Nation of Home Owners'*, Conference, Sheffield City Polytechnic, April.
Hanmer, J. and Saunders, S. (1984) *Well-founded Fear*, London: Hutchinson.
Hough, M. and Mayhew, P. (eds) (1983) *Crime and Public Housing*, Home Office Research and Planning Unit: HMSO.
Hunt, P. (1989) 'Gender and the construction of home life', in G. Allan and G. Crow (eds) *Home and Family: Creating the Domestic Sphere*, Basingstoke: Macmillan.
Jamie, K. (1991) 'Wee Wifey', *New Statesman & Society*, 19 July, p. 39.
Kaluzynska, E. (1980) 'Wiping the floor with theory – a survey of writings on housework', *Feminist Review*, 6: 27–54.
Kerouac, J. (1958) *On the Road*, London: Deutsch.
Madigan, R., Munro, M. and Smith, S. (1990) 'Gender and the meaning of home', *International Journal of Urban and Regional Research*, Vol. 14, No. 4: 625–47.
Mason, J. (1989) 'Reconstructing the public and private: the home and marriage in later life', in G. Allan and G. Crow (eds) *Home and Family: Creating the Domestic Sphere*, Basingstoke: Macmillan.

Matrix (1984) *Making Space: Women and the Man-made Environment*, London: Pluto Press.
Morris, J. (1991–2) '"Us" and "them"? Feminism research, community care and disability', *Critical Social Policy*, 33.
Morris, J. and Winn, M. (1990) *Housing and Social Inequality*, London: Hilary Shipman.
Oakley, A. (1974) *The Sociology of Housework*, London: Martin Robertson.
Pahl, R. E. (1984) *Divisions of Labour*, Oxford: Basil Blackwell.
Phillips, A. and Taylor, B. (1980) 'Sex and skill: notes towards a feminist economics', *Feminist Review*, 6: 79–88.
Pirsig, R. (1976) *Zen and the Art of Motorcycle Maintenance*, London: Corgi.
Pugh, C. (1990) 'A new approach to housing theory: sex, gender and the domestic economy', *Housing Studies*, Vol. 5, No. 2: 112–29.
Rapoport, R. and Rapoport, R. (1971) *Dual Career Families*, Harmondsworth: Penguin Books.
Resource Information Service (1988) *Woman's Housing Handbook*.
Roberts, M. (1989) 'Designing the home: domestic architecture and domestic life', in G. Allan and G. Crow (eds) *Home and Family: Creating the Domestic Sphere*, Basingstoke: Macmillan.
Saunders, P. (1990) *A Nation of Home Owners*, London: Unwin Hyman.
Saunders, P. and Williams, P. (1988) 'The constitution of the home: towards a research agenda', *Housing Studies*, Vol. 3, No. 2: 81–93.
Sexty, C. (1990) *Women Losing Out: Access to Housing in Britain Today*, London: Shelter.
Somerville, P. (1989) 'Home sweet home: a critical comment on Saunders and Williams', *Housing Studies*, Vol. 4, No. 2: 113–18.
Stanley, L. and Wise, S. (1983) *Breaking Out: Feminist Consciousness and Feminist Research*, London: Routledge & Kegan Paul.
Stubbs, C. (1988) 'Property rights and relations: the purchase of council housing', *Housing Studies*, Vol. 3, No. 3: 145–58.
Watson, S. (1988) *Accommodating Inequality: Gender and Housing*, Sydney: Allen & Unwin.
Wilson, E. (1980) *Only Halfway to Paradise: Women in Postwar Britain: 1945–1968*, London: Tavistock Publications.
Woolf, V. (1929) *A Room of One's Own*, London: The Hogarth Press.

Chapter 3

Women and owner occupation in Britain
First the prince, then the palace?

Rose Gilroy

OWNER OCCUPATION: MORE THAN JUST A TENURE

Status quo views suggest, then, that owner occupation acts as if it were a fairy godmother's wand. When waved, the wand transforms an ordinary pumpkin and insignificant white mice (a rented house) into a glittering coach and horses (a home of one's own). Similarly, a previously ragged and unhappy Cinderella (a tenant) is changed into a beautiful and desirable person (an owner occupier) who can then join with similar fortunate people in a nationwide property owned democracy in all the fun and joy of the ball.

(Merret and Gray, 1982)

Consider the image of Cinderella. It might be argued she is an archetypal woman: the uncomplaining domestic slave who through passivity and personal beauty wins her prince, her palace and everlasting happiness. To what extent is this true of the British woman in the 1990s? Is she still obliged to enter owner occupation via the church or the registry office door?

Data from the 1991 census reveal that in England and Wales there were a total of 19,877,272 households[1] of which 69.4 per cent were headed by men and the remaining 30.54 per cent by women.[2]

Table 3.1 gives head of household by gender but it is important to note that like is not being compared with like. Where women identify themselves as head of household they will generally be in households without men. Custom and practice identify the man as head of household where there is a man and a woman. The 1991 Census advice was that the person in the household who was responsible for budget might be designated as head of household

Table 3.1   Tenure of male and female heads of household (as a percentage of each tenure category)

	Women %	Men %
Owner occupation	23.81	76.19
Private rented	30.37	69.63
Housing association	53.78	46.22
Local authority	47.46	52.54
All tenures	30.54	69.46

Source: Census, 1991

where there was doubt among respondents. Whether this advice altered any predetermined thoughts is, of course, impossible to quantify.

Women heads of household will generally be single women (not cohabiting, separated, divorced or widowed), single mothers, lesbian households, women caring for dependent adults. Men identified as head of household are not necessarily going to be single men, single fathers or carers: the majority will be in families with either a non-waged partner or a working partner giving them greater power in a market-oriented system. This will be examined later in the chapter in considering women's level of economic activity and wage levels.

In fact, data from the General Household Survey 1991 reveal that a far higher percentage (77 per cent) of married couples are buying or own their own homes than any category of single men or women. See Table 3.2. Marriage, as opposed to cohabitation, is also a factor: a 1993 study revealed a higher incidence of local authority renting among cohabiting couples with children compared with married couples with children (Kiernan and Estaugh, 1993).

Table 3.2   Home owners by sex and marital status

Married couples		77%
Women	single	44%
	widowed	51%
	divorced/separated	46%
Men	single	54%
	widowed	50%
	divorced/separated	54%

Source: Table 3.33b, General Household Survey, 1991

Taking these two tables together it can be clearly seen that women score poorly in tenures where ability to pay is the entry requirement. This is confirmed by Table 3.3 which shows male and female headed households in rented dwellings: both the public sector accessed by conforming to assessments of housing need and the private rented sector where ability to pay is a determinant of access and quality.

Table 3.3 Tenants by sex and marital status

	LA/HA/ Coop %	Private renting		
		with job %	unfurn. %	furn. %
Men				
single	26	3	6	11
divorced/separated	32	2	6	5
widower	43	1	6	1
Women				
single	43	1	6	6
divorced/separated	48	1	4	2
widow	43	0	5	0

Source: Table 3.33b, General Household Survey, 1991

Men are more likely to be in private renting than women and, significantly, are more likely to be found in furnished lettings. While it is difficult to generalise about the private rented sector, given its heterogeneous nature, it remains true that furnished lettings usually exact higher rents. These, of course, may be paid by Housing Benefit depending on local Housing Benefit ceilings; however, access to this tenure type is usually by substantial deposits and rent in advance, neither of which is available from the State. From this it may be surmised that divorced and separated men and single men have higher disposable incomes than women in these household groups.

The most obvious discrepancy is that women are trailing badly behind in owner occupation. This is a cause for concern: with the growing ideological commitment to home ownership and the concomitant reduction in the capital and revenue support for social housing, owner occupation has become the dominant tenure. Not

Table 3.4   Views of women on home ownership

	Accrington	Consett	Cheltenham	All
Sample	200	200	203	603

(a) People who are successful in life become home owners:

	Accrington	Consett	Cheltenham	All
Strongly agree/ tend to agree	63	66	64	65
Neither agree/ disagree	9	5	7	7
Strongly disagree/ tend to disagree	23	27	28	26
No opinion	4	–	2	2

(b) People naturally prefer to own their own homes:

	Accrington	Consett	Cheltenham	All
Strongly agree/ tend to agree	80	82	78	80
Neither agree/ disagree	5	4	7	6
Strongly disagree/ tend to disagree	12	14	14	13
No opinion	3	1	1	2

only the dominant tenure in terms of numbers of households it encompasses, but psychologically owner occupation is seen as a mark of success. This is witnessed by a number of surveys including the following which conveys the views of women in three areas: Accrington, Cheltenham and Consett (Forrest and Murie, 1986).

Table 3.4 shows that views in Consett were the most polarised and it is worth considering that in 1980 Consett, a small town in County Durham, was a victim of the restructuring of the British Steel Corporation and passed from BSC to MSC almost overnight.

In spite of their plunge into recession and the experience of home owners finding themselves with an unsaleable asset, the view of owner occupation is still an optimistic one. This may suggest that central government promotion of owner occupation has successfully raised owner occupation into a Hobson's choice tenure, with local authority renting transformed into a shameful demonstration of dependency. With owner occupation associated with success, what then is the position of women?

The British form of owner occupation, with mortgages which bear down heavily in the early years, favours high or dual earners and therefore generally not women.

Similarly until recently the escalating price of dwellings was also a dominant factor. In such a market the poorer earning capacity of women, exemplified by lower earnings, a career likely to be broken by childcare, by caring for a dependent adult, militates against a woman gaining sufficient financial independence to purchase and retain an owner-occupied dwelling alone.

This raises the question of whether new forms of ownership, such as equity sharing, discounted house purchase brought about by the right-to-buy or do-it-yourself home ownership, bring new opportunities for women. These issues will be discussed later in the chapter.

Beyond these economic reasons Watson (1986; 1988) and McCarthy and Simpson (1991) detect an ideological link between owner occupation and the domestic cycle of the nuclear family: 'with economic notions of stability, accumulation, appreciation and inheritance representing and reflecting the formation, growth and reproduction of an idealised domestic unit' (McCarthy and Simpson, 1991: 27). The idealised domestic unit has long been viewed as:

> the young, married, heterosexual, white middle-class couple with two children, a boy (older) and a girl, all of whom live together in their own house. The husband is the main bread-winner and the wife is a full-time housewife/mother who may, however, work part time.
>
> (Gittins, 1993: 3)

The assumption of this ideal as the norm can be seen to influence social policy, including housing, as witnessed by the now famous quotation from the 1971 White Paper on home ownership:

> Home ownership is the most rewarding form of housing tenure. It satisfies a deep and natural desire on the part of the house-holder to have independent control of the home that shelters *him* and *his* family. It gives *him* the greatest possible security against the loss of *his* home; and particularly against the price changes that may threaten *his* ability to keep it. If the house-holder buys his house on mortgage *he* builds up by steady saving a capital asset for *himself* and *his* dependents.
>
> (DOE, 1971; my italics)

Home ownership by women is usually concomitant with marriage but, in addition, there have always been:

three socially acceptable models for home ownership by women: separated and divorced women usually retained the family home and stayed there to raise the children; widows lived their lives out in their married locations; and single 'spinsters' might inherit a home from their parents.

(Card, 1980: s. 216)

Women's home ownership, then, is generally one of ownership and not purchase and that ownership has been arrived at through their relationship with a man. The next section looks at these three acceptable models.

Cinderella, we are encouraged to believe, lived happily ever after but for many women marital happiness is followed by marriage breakdown. What happens to divorced and separated women?

THE DIVORCEE AND THE SINGLE PARENT

With owner occupation the most common tenure for married couples, the rising divorce rate has made possession of the marital home a critical issue. Card's analysis and common mythology states that women retain the marital home. This seems to be borne out by the findings of the General Household Survey, 1991 which are set out in Table 3.5.

According to these data women have a better chance of retaining their status as owner occupiers than men in the long and short term, while men have a greater chance of improving their housing status from tenants to owner occupiers. These data simply analyse gender, but a closer analysis based on gender and custodial status of parents by McCarthy and Simpson (1991) presents a different view of movers and stayers.

Custodial fathers, as Table 3.6 shows, have a greater chance of remaining in the marital home than custodial mothers. Of particular significance is the greater staying power of custodial fathers in owner occupation.

Initially, more than 90 per cent of custodial fathers remained in the marital home at separation, though a third moved out later making a total movers group of 39 per cent. This compares with 62 per cent of custodial mothers who remained in the marital home, of which one-third moved making a total of 59 per cent of custodial mothers who moved at some point.

Table 3.5 Summary of tenure status of the marital home; one year after divorce and at the time of interview

Tenure of the former marital home	Tenure one year after divorce	Current tenure	Men %	Women %	Total %
Owner occupied	Owner occupied	Owner occupied	33	36	35
Owner occupied	Rented/not a householder	Owner occupied	6	7	7
Owner occupied	Owner occupied	Rented/not a householder	3	3	3
Owner occupied	Rented/not a householder	Rented/not a householder	17	13	14
Rented	Rented	Rented	18	27	23
Rented	Owner occupied/not a householder	Rented	3	3	3
Rented	Rented	Owner occupied/not a householder	8	6	7
Rented	Owner occupied/not a householder	Owner occupied/not a householder	12	6	8
Base = 100%			204	339	543

Source: Table 3A, General Household Survey, 1991

Table 3.6 Decisions over the marital home by gender, custodial arrangements and tenure of marital home

	Stayed in home	Moved later	Moved mutual	Moved first	All cases
	No %	No %	No %	No %	No %
Custodial fathers					
Owner occupied	39(62)	20(32)	1(2)	3(5)	63(100)
Council	4(50)	2(25)	0(0)	2(25)	8(100)
Other	3(75)	1(25)	0(0)	0(0)	4(100)
All tenures	46(61)	23(31)	1(1)	5(7)	75(100)
Custodial mothers					
Owner occupied	113(44)	52(20)	29(11)	63(25)	257(100)
Council	34(44)	11(14)	3(4)	29(38)	77(100)
Other	4(13)	14(45)	4(13)	9(29)	31(100)
All tenures	151(41)	77(21)	36(10)	101(28)	365(100)
Non-custodial fathers					
Owner occupied	18(13)	15(11)	13(10)	90(66)	136(100)
Council	5(22)	1(04)	2(9)	15(65)	23(100)
Other	2(10)	5(24)	5(24)	9(43)	21(100)
All tenures	25(14)	21(12)	19(11)	114(64)	180(100)
Non-custodial mothers					
Owner occupied	0(0)	1(6)	1(6)	16(88)	18(100)
Council	0(0)	0(0)	0(0)	3(100)	3(100)
Other	0(0)	0(0)	0(0)	1(100)	1(100)
All tenures	0(0)	1(5)	1(5)	20(90)	22(100)

Note: Percentages in tables may not total 100 per cent because of rounding error.

Source: McCarthy and Simpson, 1991; reproduced by permission of Avebury Books, Ashgate Publishing Ltd

The group least likely to remain in the marital home were non-custodial mothers. In the McCarthy and Simpson (1991) study 21 out of 22 women in this group had moved out at the time of separation while the remaining woman subsequently moved into the home of her new partner.

Non-custodial fathers were highly mobile with 87 per cent moving at some point, though 20 per cent initially remained in the marital home including 13 per cent who remained in owner occupation.

Of greater significance is not who remains initially in the marital home but who has the ability to retain it. While custody of children seems to have been the critical issue determining whether a man remained in the owner-occupied marital home, the critical issue for women was income levels. Table 3.7 indicates the relationship between occupational status for women and tenure change following marriage breakdown.

Interestingly, women in the higher earnings bracket (professional and managerial) not only retained owner occupation status but were more likely to increase their take-up of this tenure, in spite of the fact that a significant number initially moved into rented property or shared with friends or family.

Men in this higher earnings group fared less well (see Table 3.8). Significant numbers moved initially into private renting and overall the number remaining in owner occupation was slightly less than those who were owners prior to divorce.

Similarly, comparing men and women in intermediate occupations, women increased their take-up of owner occupation, while a significantly higher proportion of men lost out going to a shared solution or to private renting.

In the lower level group of semi-skilled and unskilled workers both men and women lost ground in owner occupation and increased their reliance on council housing, revealing the importance of two incomes for the sustaining of owner occupation to those in lower paid occupations. Given the poorer wages of women in this sector particularly, it is not surprising to see women losing more ground in owner occupation and increasing their take-up of council tenancies by 20 per cent compared to an increase of 8 per cent by men.

With regard to men and women without waged work at the time of divorce, more women lost ground with 17 per cent dropping out of owner occupation compared to only 9 per cent of men. The reasons for women's greater movement out of owner occupation are summed up by Millar (1988) who declares that the period after separation or divorce is 'characterised by financial hardship when financial needs and financial resources are likely to be very far apart' (p. 100).

McCarthy and Simpson's research (1991) reveals the particularly invidious position suffered by those women whose marriages break down within ten years. These women are likely to be mothers of young children which increases the likelihood of the women being unwaged at the time of divorce and decreases their ability to take up employment, given the availability and cost of

Table 3.7 Occupational status and tenure changes of women

	Occupational status					
	Owner occupier %	Council rent %	Private rent %	Other rent %	Shared %	No. (100%)
Higher						
Marital tenure	89	4	4	4	0	82
First move	57	2	10	6	25	49
Current tenure	94	4	1	1	0	82
Intermediate						
Marital tenure	75	17	5	3	0	65
First move	34	11	11	6	37	35
Current tenure	79	12	2	5	3	65
Lower						
Marital tenure	73	19	4	4	0	26
First move	14	14	21	14	36	14
Current tenure	54	39	4	4	0	26
Not working						
Marital tenure	63	30	5	2	0	184
First move	17	19	10	12	41	115
Current tenure	46	39	4	7	4	184

Source: McCarthy and Simpson, 1991; reproduced by permission of Avebury Books, Ashgate Publishing Ltd

Table 3.8 Occupational status and tenure changes of men

	Occupational status					
	Owner occupier %	Council rent %	Private rent %	Other rent %	Shared %	No. (100%)
Higher						
Marital tenure	92	1	5	2	0	86
First move	38	2	38	4	18	55
Current tenure	86	4	9	1	0	86
Intermediate						
Marital tenure	83	7	6	4	0	81
First move	27	2	21	4	46	56
Current tenure	70	7	12	4	6	81
Lower						
Marital tenure	70	22	9	0	0	23
First move	14	21	14	7	43	14
Current tenure	52	30	4	4	9	23
Not working						
Marital tenure	50	44	6	0	0	32
First move	21	17	24	0	38	29
Current tenure	41	47	0	0	12	32

Source: McCarthy and Simpson, 1991; reproduced by permission of Avebury Books, Ashgate Publishing Ltd

childcare. A woman in such a position is likely to be in a property at the lower end of the market in which, because of the short duration of her marriage, she is likely to have little equity. This combination of events will mean that, on resale of the property, such a woman will have few assets to enable her to trade down, supposing there is any property in a lower price bracket. A new problem facing divorcing mothers is the Child Support Act, which may put an end to the common agreement that a woman will forgo maintenance in exchange for the marital home. In the future, maintenance may be compulsory, reducing the likelihood of a husband giving up his home and increasing his financial interest in it. This may well reduce a woman's ability to retain the marital home and increase her reliance on the rented sector through homelessness at the point of separation, as opposed to homelessness brought about by mortgage arrears and repossession.

An issue for the future is the rise in the numbers of divorced women among the elderly. On divorce a woman will lose the future security of her husband's occupational pension. If her own employment history is punctuated by career breaks she may find that her entitlement leaves her in poverty.

THE WIDOW

The tables from the General Household Survey indicate that the widow outstrips her sisters in respect of owner occupation and is marginally more likely to be in owner occupation than the widower.

The privileged position of widows is due to demographic trends which show that women marry younger than men, tend to marry men older than themselves and live longer, thus increasing their chances of becoming widows. The problem for this group, which is discussed more fully in the chapter on elderly women by Roger Sykes (Chapter 5), is that many widows fall into the category of not rich, not poor and find themselves denied assistance from the State, though their own funds may be insufficient to maintain the property in terms of decoration and, more critically, repair and adaptation.

THE CARING DAUGHTER

With the rapid growth of home ownership and the rise (until recently) of house prices, housing inheritance is likely to become a more important issue.

The growth in home ownership has not yet trickled down into housing inheritance, though there has been a growth of some 20 per cent in the number of estates containing residential property over the last 25 years. According to a study published in 1991 (Hamnett and Williams), about 1 per cent of the adult population inherit a property every year with a total of some 8 per cent or 3.7 million adults having inherited a residential property or a share of the proceeds.

It is not surprising to note that the spread of inheritance is uneven and dependent on class, age, tenure and region. Housing inheritance is more common in the South East where owner occupation has been established longest. The greatest number of inheritors were those aged over 50. Those most likely to inherit were from social class A and those least likely to inherit were from the other end of the class spectrum. Similarly, it comes as no surprise to note that those who are already home owners are more likely to inherit (17 per cent) compared to only 3 per cent of council tenants. What is the position of women in this? The research shows that 75 per cent of inherited properties were immediately sold, while 25 per cent were retained, generally by a beneficiary who was resident in the property. In the majority of cases the beneficiary is the widow and only in a very few cases are we looking at the 'caring daughter'.

Current movements in social policy create concerns for feminists. Central government's policy of community care is only a thin disguise for the reality of care by female family members.

Data from the General Household Survey (1992) reveal that 13 per cent of men and 17 per cent of women are caring for someone who is sick, elderly or disabled. The greater numbers of women in the population suggest that the true number of women carers exceeds the number of male carers with figures of 2.9 million men compared to 3.9 million women.

Similar proportions of men and women care for someone in the same household but more women care for a friend, neighbour or relative in another dwelling.

The age group of carers is more likely to be the 45–64 band when parents can be expected to show increasing frailty. As might be anticipated, single women are more likely to be carers than single men, though both are more likely caring for parents than their married, divorced or widowed brothers and sisters. The married sister or brother is more likely to have an owner-occupied home of their own, as are the divorced or widowed siblings. What happens to the unmarried daughter who has come back to care?

What is not available is the tenure of those who are providing care or the tenure of the dependent adult, and therefore it is impossible to determine how many women carers may inherit the property of the dependant. However, what is fairly clear is that the ability of such women to inherit property in the future may be severely limited by pressure on the ageing population. As longevity increases and the cost of residential or nursing care falls increasingly on individuals, those who are home owners may need to release equity to fund open-ended care arrangements if they become too frail to be cared for at home. At the time of writing levels of concern are rising as to the long-term prospects of the

Table 3.9 Percentage of adults aged 45–64 caring for family members, other relatives and friends, by sex and marital status

	Marital status			
	Married/ cohabiting	Single	Widowed/ divorced/ separated	Total
Percentage of men aged 45–64 who were caring for				
Spouse	2	–	–	2
Child 16 and over	1	0	0	1
Parent	9	16	11	9
Parent-in-law	7	–	0	6
Other relative	3	3	2	3
Friend or neighbour	4	8	3	4
Base = 100%	2024	149	216	2389
Percentage of women aged 45–64 who were caring for				
Spouse	3	–	–	2
Child 16 and over	1	0	1	1
Parent	15	20	11	15
Parent-in-law	5	0	0	4
Other relative	4	12	4	4
Friend or neighbour	7	15	9	8
Base = 100%	1982	126	490	2598

Source: Table 8, General Household Survey: carers in 1990

state pension. Might it be entirely phased out or retained only for those who have been unable to buy into private pension schemes? Far from being a means of transferring wealth to future generations, the ownership of a property may be a savings scheme against future frailty and poverty.

So far this chapter has examined the tenure position of women by determining their marital status. The next section focuses on the economic position of women-headed households.

POOR WAGES, POOR HOUSING

The relative position of women heads of household in terms of economic status is shown in Table 3.10. While two-thirds of male heads of household are economically active, the greatest number of women-headed household are those who are inactive. The idea of economic inactivity is interesting. The category is made up of students, the permanently sick, the retired – a group dominated by women and 'other inactive'. This final group is composed of those with home responsibilities, i.e. housewives, stay-at-home mothers and carers of dependent adults: in short, women. Figures from the Labour Force Survey 1991 reveal that economically inactive women make up 29 per cent of that female adult population compared to 12 per cent of men in this category (*Employment Gazette*, 1992). The difference can be accounted for partly by the greater longevity of women, making them the biggest group in the retired category, but also by the 58 per cent of inactive women who said they were caring for family and home – a role which men less commonly take. The figures show an increasing participation by women in the labour market with a decrease of 17 per cent in the numbers of inactive women from 1979, while the corresponding figures for men have increased by 51 per cent reflecting early retirement and the shedding of older men in the workforce. While more women are in waged work, an analysis of their tenure pattern shows them firmly rooted in rented housing. See Table 3.11.

Economically inactive women make a strong showing in all tenure groups, forming a majority in all rented sectors. Their dominance of the housing association sector, in particular, is perhaps explained by the high proportion of sheltered housing schemes with their resident population of elderly single women.

Unemployed women figure prominently in private renting where they make up half the number of householders. Given that

Table 3.10 Economic position[3] of men and women heads of household (all tenures)

	Men %	Women %
Economically active	66.32	34.73
Unemployed	6.24	3.54
Economically inactive	27.42	61.72

Source: Census, 1991

Table 3.11 Women heads of household by tenure and economic position

	Economically active %	Unemployed %	Economically inactive %
Owner	15.57	14.27	42.17
Private rent	26.78	49.30	54.77
Housing association	37.13	30.41	64.24
Local authority	29.08	22.83	59.40

Source: Census, 1991

this sector is also accessed by ability to pay, it is important to note that the private rented sector is not homogeneous and, as with any sector operating on market forces, a tenant will get what he or she pays for and much will depend on the local ceilings on rent levels imposed by Housing Benefit officers.

It is in the category of economic activity that the position is markedly different from that of male-headed households. Michael Ball (1982) has stated that: 'owner occupation is now the majority tenure for the economically active sectors of the working class (on virtually any definition of that class).' In the category of economically active heads of household who are owner occupiers only 15.57 per cent are women. This poor showing demonstrates that Ball was focusing on the nuclear family in his analysis.

One of the issues is that the category of economically active embraces both full-time and part-time workers but evidence from the Labour Force Survey, 1991 (*Employment Gazette*, 1992) shows that 42 per cent of women in waged work are working part time. See Table 3.12.

Table 3.12   Employment by sex and full-time or part-time work in the UK
(figures in thousands)

	Men		Women	
	Full time	Part time	Full time	Part time
1991	13,274	603	6068	4482

Source: Labour Force Survey, 1991

Table 3.13   Real[4] weekly earnings in 1991 after tax,
National Insurance, Child Benefit and Family Credit

	Single man	Single woman
Lowest decile point	125.5	99.20
Median	202.6	148.60
Highest decile point	365.2	252.70

Source: London Housing Unit, 1993b

For those in full-time work there is a marked discrepancy between the earnings of men and women. See Table 3.13. Wage
levels vary around the country as, of course, do opportunities for
employment. Do these make a difference to the housing opportunities of women-headed households?

Data from the 1991 census reveal interesting spatial differences.
While economically active women make up only a small number of
owner occupiers, an economically active woman has the greatest
chance of being a home owner if she lives in Inner London. Given
the acceleration of house prices in the capital this is perhaps
surprising. It could be expected that women might do better in the
North East where house prices are traditionally lower. However,
regardless of the ratio of incomes to house prices there is in the
North a substantial proportion of households, particularly women-
headed households, where the income level is such that owner
occupation is not a realistic option (Regional Equity Group,
1992). A similar picture is painted in Wales (Shelter Cymru, 1990).
The London position is a reflection of the concentration of high
earning women. See Table 3.14.

This is borne out by an analysis by Michael Harloe (1992) of the
housing circumstances of Londoners which revealed that in the

Table 3.14 Percentage of women-headed house-holds of all households by tenure type and district type

District type	Economically active owners %
Inner London boroughs	30.81
Outer London boroughs	19.34
Principal metropolitan cities	17.73
Other metropolitan districts	14.34
Large non-metropolitan cities	16.95
Small non-metropolitan cities	18.81
Districts with industrial areas	13.18
Districts with new towns	13.94
Resort ports and retirement districts	16.47
Urban and mixed urban/rural districts	13.91
Remoter, mainly rural areas	13.17

Source: Census 1991

small number (38 or 8.4 per cent) of women-headed households, where the woman was in a high status and better paid professional or managerial job, these women were slightly more likely than their male counterparts to be buying their own home. Significantly, all other groups of women-headed households, including those in waged work, had less access to home ownership than men.

Recent work undertaken by the London Housing Unit (1993b) indicates the affordability gap suffered by most women. The figures seem to show that 54 per cent of full-time working women could afford to put their foot on the London property-owning ladder but, given the high proportion of women who work part time and those who are economically inactive, the percentage of adult women who can purchase in London is reduced to only 15 per cent. See Table 3.15.

Prior to the passing of the American Equal Credit Opportunity Act (1973), and the British Sex Discrimination Act 1975, discriminatory practice by lenders was an element in blocking women's access to mortgage facilities (Merret and Gray, 1982). Since this has been made unlawful it is clear that the action of gatekeepers was only part of women's exclusion from the sector. A greater part has been, and is, played by women's poorer wages and their

Table 3.15 Income needed to buy a property in lowest decile band (Greater London, 1992)

	1 bed	2 bed	3 bed	4+ bed
1 earner:				
£ per year	13,320	16,200	19,800	28,080
£ per week	256	312	381	540
Percentage of full-time workers earning this amount				
M	73%	57%	41%	18%
F	54%	32%	17%	NA

Source: London Housing Unit, 1993b

limited access to full-time waged work given their domestic and caring responsibilities.

Recent research by Peach and Byron (1993), exploring the reasons for the high concentration of Caribbean households in council rented property, has rejected the earlier class-based explanations and highlights the significance in housing opportunity terms of the high proportion of women-headed households.

WOMEN AND NEW OWNERSHIP INITIATIVES

The Conservative government has increased opportunity for home ownership by encouraging different mortgage forms and part-rent/part-purchase schemes. The most successful scheme, however, has been the introduction of the Right to Buy in 1980. Given the large numbers of women-headed households in the social rented sector, this initiative, which lowers the purchase price, could be viewed as giving a boost to the chances of women with lower earnings becoming home owners. Evidence from the North East suggests otherwise.

The Nationwide Anglia Building Society's (1990) analysis of women purchasers revealed that in the North nearly one-third (28.6 per cent) of Right-to-Buy purchasers were women in 1989 compared to almost a quarter (20.9 per cent) of other first-time buyers and 20.9 per cent of all purchasers except council tenants. Finer grained data from the Halifax Building Society of purchasers between 1985 and 1988 reveal smaller numbers of women buyers as well as interesting details of their household groupings. See Table 3.16.

Table 3.16 Right-to-Buy purchasers by household type and income group (1985–8) in the Newcastle travel to work area

Household type	Income group (£)								Total
	Up to 4999	5000– 7499	7500– 9999	10,000– 12,499	12,500– 14,499	15,000– 17,499	17,500– 19,999	20,000–	
YSP	0	0	0	1	0	0	0	0	1
DINK	8	33	38	29	11	8	3	0	130
YFAM	0	0	0	0	0	0	0	0	0
MATFAM	2	8	5	5	4	1	1	0	26
MAC	12	83	129	96	61	16	4	4	405
SMA	4	17	7	5	0	0	1	0	34
SPF	7	12	8	3	1	1	0	0	32
OAPC	14	11	4	3	1	0	0	1	34
SOAP	22	3	1	1	1	0	0	0	28
Total	69	167	192	143	79	26	9	5	690

Source: Robinson-Lundy, 1990

Note: Young (under 25) single person (YSP); young childless couple (DINK); couple plus 1–2 children under 16 (YFAM); couple plus 3 or more children (MATFAM); middle-aged (25–60) couple (MAC); single middle-aged (SMA); single parent family (SPF); elderly (60 plus) couple (OAPC); single elderly (SOAP).

The data show that two household groups dominate the group of purchasers: young childless couples and middle-aged couples. The latter is to be expected, given that these couples probably no longer have the expense of childcare and may have built up sizeable discount through long tenancy. The young childless couple is more surprising, but this is a dual-income couple without the expense of child-rearing, though this will be matched by poorer discounts. Both these groups most commonly had an income of between £7500 and £9999 although the income of the older couples was skewed slightly higher. These findings are substantiated by the General Household Survey, 1991 which shows married couples benefiting particularly from the scheme.

When compared to other owner occupiers (non-Right to Buy) the groups which seem to have gained from the scheme are pensioner couple households and large adult households. In the latter purchasers may benefit from the ability to name other purchasers and therefore include their earnings in their mortgage eligibility. In the case of pensioner couples it may be that sons and daughters are purchasing their parents' homes, thinking ahead to their own asset accumulation or, indeed, depending on location, to a second home.

Household groups in which women might figure as head of household, that is, the single elderly person and the single parent family, account for very few purchasers (13 per cent in total of the Halifax purchasers). All these fall into the lowest annual income brackets. See Table 3.17.

Ironically, when Lady Thatcher, then the Prime Minister, was promoting the benefits of her soon to be introduced Right-to-Buy scheme, the image she chose was this:

> Mothers with small children living in tower blocks, just as anyone else living in tower blocks, will, under a Conservative Government, now have three options: to carry on renting, to put down an option to purchase the flat within a reasonable time, or to purchase the flat. That seems to me to enlarge the freedom and possibilities available to such people.
>
> (House of Commons, *Parliamentary Debates*, 1979–1980)

For such a woman, perhaps coping on low wages, any initiative designed to help her to become a home owner may well be misdirected. Christine Whitehead (1986) summarises her position in a common-sense way:

Table 3.17 Comparison of heads of household who have bought local authority accommodation with heads of household in owner-occupied and local authority accommodation: selected characteristics

	Bought LA accommodation %	Not bought LA accommodation	
		Owner occupier %	Rents from local authority %
Age			
16–19	2	11	15
30–39	12	21	14
40–49	24	21	11
50–59	25	16	13
60–64	11	8	7
65–69	11	7	11
70–74	10	5	9
75 and over	5	10	19
Average (mean) age	54	50	54
Base = 100%	805	5839	2313
Marital status			
Married	73	71	43
Single	4	9	15
Widowed	14	12	16
Divorced/separated	9	7	16
Base = 100%	805	5839	2313
Household type			
1 adult aged 16–59	5	9	10
2 adults aged 16–59	16	19	8
Small family	13	21	20
Large family	5	6	9
Large adult household	28	16	11
2 adults, 1 or both aged 60 or over	22	16	16
1 adult aged 60 or over	11	12	27
Base = 100%	805	5839	2313

Source: Table 3.7, General Household Survey, 1991

the ownership of a housing asset is not necessarily the best use of limited resources for low income households. Such households may rather spend what little they have on other, usually consumer goods. Secondly, the benefits of owner occupation remain highly correlated with income and wealth. Those at the bottom of the scale often do not benefit from freedom and security and may in some circumstances even suffer capital losses.

(Whitehead, 1986: 74)

Factors other than poverty may also be operating. Research currently being undertaken by the University of Newcastle is examining whether women are being discriminated against in public sector allocations. While findings have yet to be published, it is credible that officer judgements about class, respectability and their own ideas of the family may operate against certain groups, including the single mother of three children, in terms of allocating quality property. Women who are allocated poorer quality property may decide, whatever their purchasing potential, that their 'Right to Buy' is an offer they can refuse.

For a woman with a moderate income and/or some equity from the sale of her parents' home or her former marital home, an initiative such as shared ownership seems a useful means of helping her to enter or re-enter owner occupation. Shelter Cymru's survey (1990) found a high number of satisfied women who were occupying properties bought under these arrangements. The Welsh survey apart, current research (Cousins et al., 1993) conflates the position of men and women giving no insights into the extent to which women might be benefiting (or not) from this initiative. The report reveals that shared owners are typically childless couples (37 per cent) or single people (25 per cent) and younger (65 per cent under 30 years of age) than heads of household nationally. Compared to all first-time buyers, shared ownership purchasers had lower incomes than the median income of single people (£160–£200) and single parents lower again at £120–£160. Again, the research states that 14 per cent of shared ownership purchasers were formerly in owner occupation, of which 27 per cent were divorced or separated. These facts together seem to point to women being natural beneficiaries of shared ownership schemes. Whether they are is not revealed: purchasers are analysed by income, socio-economic group, age, household style and ethnicity, but frustratingly not by gender.

Other private sector schemes which offer 'discount purchase' offer a buy now, worry later package (Norden, 1993). These schemes offer the purchaser 50 per cent of the property with a commitment to buy the other half five years later. Such an incentive is built on the concept that within the five-year period the purchaser will increase his or her earnings. Norden's investigation tells tales of woe about a young married couple and a single man. How much more at risk might a single woman be, given that she might find herself with lower wages and making slower progress in her career than a man? Similar warnings extend to schemes involving deferred or index-linked payments which again presuppose increased prosperity. A single woman may not enjoy the same accelerated career as a man and a woman with children may have a long-term commitment to paying for childcare in order that she may work.

Attaining a mortgage, then, may not be the beginning of the happy ending so much as the start of financial misery. Considerable attention has been paid in the last three years to the problems of those with negative equity and those facing homelessness because of rising mortgage interest payments, redundancy and falling values. In all this the position of women remains unclear for data available from the county courts on possession do not record gender. The traditional view that arrears are linked to the 'four Ds' – divorce, death, disability and the dole – suggests an ominous position for women, given their greater likelihood of being widowed, of becoming disabled, and the impact of divorce particularly on women with children. Other evidence from the Waltham Forest Advice Centre (LHU, 1993a) reveals that women make up 54 per cent of the Centre's clients from all tenures with debt problems, while 71 per cent of council tenants who go to the Centre with rent payment problems are women.

CONCLUSION

This chapter has concentrated on current data to illustrate that, in spite of the changing shape of British society and the family, the position of women in terms of owner occupation is still one where the woman has to find her prince before she gets her palace. For the woman who is left a widow or who retains the marital home after divorce there may be a struggle either to pay the mortgage and/or maintain the property. A woman who may have interrupted her career to care for dependent adults may face the same financial problems. All these are compounded by the poor attention that women receive in building up their DIY skills and their related lower confidence levels in tackling jobs of this kind.

This discussion has not answered the question about what quality of property women buy. While access *per se* is still an issue, access to quality and safe property is also important. The Nationwide Anglia report *Lending to Women* (1989) stated that women bought cheaper property with poorer amenities – no central heating, no garage – and older properties with therefore greater potential maintenance costs. Since 1988 there has been no further study to draw upon and plainly this is an area which requires research in order that the position of women is adequately reflected in any changes made to the grant regime.

If women are to be helped into owner occupation, an effective mechanism needs to be found, perhaps part rent/part buy though evidence is scant. A necessary part of any package is a means of 'staircasing down' which could be introduced at times of hardship such as those characterised by the 'four Ds'. What is needed is more funding for equity share schemes and for mortgage rescue, both of which should be targeted at women-headed households.

While the ideological obsession with tenure continues there must be ways of increasing women's access to owner occupation, but we also need to voice the question 'What is wrong with renting?' For many women-headed households there are not, and probably never will be, the income levels to support owner occupation adequately. For these households there needs to be the option of responsibly managed, quality, rented property at affordable levels.

NOTES

1 For census purposes a household is defined as a person living alone, or a group of people who have the address as their only or main residence and who either share one meal a day or share the living accommodation. This definition masks the true number of households swollen by those living with parents or friends who could be described as homeless.

2 My thanks to Dan Dorling, Joseph Rowntree Fellow at the University of Newcastle for providing the data from the census.

3 Economic position is classified in census as follows:

Economically active
Persons in employment
Employees
(i)   Full time
(ii)   Part time
Self-employed
(iii)   With employees
(iv)   Without employees
(v)   On a government scheme
Unemployed
(vi)   Waiting to start a job
(vii)   Seeking work
(viii) Students (included above)
Economically inactive
(ix)   Students
(x)   Permanently sick
(xi)   Retired
(xii)   Other inactive.

4   Real earnings as at April 1991.

REFERENCES

Ball, M. (1982) 'Housing provision and the economic crisis', *Capital and Class*, 17: 63.

Card, E. (1980) 'Women, housing access and mortgage credit', *Signs*, Spring: 215–19.

Cousins, L., Ledward, C., Howe, K., Rock, G. and Taylor, G. (1993) *An Appraisal of Shared Ownership*, London: Department of the Environment, HMSO.

Department of the Environment (1971) *A Fair Deal for Housing*, Cmnd 4728, London: HMSO.

Employment Gazette (1992) 'Women and the labour market: results from the 1991 Labour Force Survey', September: 433–43.

Forrest, R. and Murie, A. (1986) Data from survey of households, carried out as part of the ESRC funded research, Urban Change and the Restructuring of Housing Provision, quoted in R. Forrest, A. Murie and P. Williams (1990) *Home Ownership: Differentiation and Fragmentation*, London: Unwin & Hyman.

Gittins, D. (1993) *The Family in Question*, 2nd edition, London: Macmillan.

Hamnett, C. and Williams, P. (1991) 'Housing inheritance in Britain', *Housing Research Findings*, Joseph Rowntree Foundation, September.

Harloe, M. (1992) 'Housing inequality and social structure in London', *Housing Studies*, Vol. 7, No. 3: 189–204.

House of Commons (1979–1980) *Parliamentary Debates*, Vol. 967, c. 1223.

Kiernan, K. and Estaugh, V. (1993) *Cohabitation, Extramarital Childbearing and Social Policy*, London: Family Policy Studies Centre.

London Housing Unit (1993a) *Collecting the Rent*, London: LHU.

London Housing Unit (1993b) *Housing the Poorer Sex*, London: LHU.

McCarthy, P. and Simpson, R. (1991) *Issues in Post Divorce Housing*, Aldershot: Gower.

Merret, S. and Gray, F. (1982) *Owner Occupation in Britain*, London: Routledge & Kegan Paul.

Millar, J. (1988) 'The costs of marital breakdown', in R. Walker and G. Parker (eds) *Money Matters: Income, Wealth and Financial Welfare*, pp. 99–114, London: Sage.

Nationwide Anglia Building Society (1989) *Lending to Women 1980–1988*.

Nationwide Anglia Building Society (1990) *Lending to Council Tenants*.

Norden, B. (1993) 'Buy now, worry later', *Roof*, May/June: 23–5.

OPCS (1992) General Household Survey: carers in 1990, OPCS Monitor, 17 November.

OPCS (1993) Labour Force Survey 1992.

OPCS (1993) Census Monitor 1991.

OPCS (1993) General Household Survey 1991.

Peach, C. and Byron, M. (1993) 'Caribbean tenants in council housing: race, class and gender', *New Community*, April: 407–23.

Regional Equity Group of Housing Associations (1992) *Regional Equity? A Policy for the Effective Regeneration of the North's Housing*, Newcastle: University of Newcastle upon Tyne.

Robinson-Lundy, S. (1990) 'An analysis of right to buy sales in the Newcastle travel to work area', Internal working paper for Joseph Rowntree Foundation Finance Project.

Shelter Cymru (1990) *Breaking Down the Barriers*, Swansea: Shelter.

Watson, S (1986) 'Housing the family: the marginalisation of non-family households in Britain', *International Journal of Urban and Regional Research*, Vol. 10, No. 2: 8–28.

Watson, S. (1988) *Accommodating Inequality: Gender and Housing*, Sydney: Allen & Unwin.

Whitehead, C. (1986), 'Low cost home ownership in the context of current government policies', in P. Booth and T. Crook (eds) *Low Cost Home Ownership*, Aldershot: Gower.

Chapter 4

Women and participation

Marianne Hood and Roberta Woods

INTRODUCTION

This chapter is about women's participation in social rented housing. It considers four main issues: what is tenant participation about; what is the need for, and value of, *collective* action by tenants and residents; what are the barriers to participation, and especially the problems that women can face; and what is to be learned from all this?

Although there has been quite a lot of research into different aspects of tenant participation, there has been very little research which has looked specifically at the importance of, and the part played by, women. It is widely known that women are not well represented in the higher ranks of policy-making bodies, and comparison is often made with the much greater extent to which women are involved in informal community and political activity. This type of involvement is more accessible than participation in formal organisations; activities are local and more flexible with, for example, the timing of meetings enabling women to combine more easily participation in community groups with domestic and waged work. Bondi and Peake (1988) suggest that women's apparent preference for participating in informal activity actually reflects women's position as workers in the domestic sphere:

> [I]n community politics there is a much greater degree of role continuity between 'political' and 'non political' activities. This is particularly important for women, for whom domestic and family roles have great salience. Moreover, the community and the home have different meanings for women than for the majority of men because, for women, regardless of their social class or ethnic origin, community and home are workplaces,

whether or not there is a workplace beyond the community. Consequently, for the housewife there is no separation in space, time or identity between 'work' and 'rest'. Thus, the home and the community as workplaces become the locus of political activity arising from issues of 'reproduction' in a manner parallel to that of the shopfloor in connection with production issues.

(Bondi and Peake, 1988: 35–6)

Women in tenants' groups are almost exclusively working class and most live in households dependent on state benefits. For financial and tax reasons, owner occupation has become the dominant way of people housing themselves in Britain, leaving in the main only those who are not in a financial position to take on a mortgage to depend on renting accommodation from a local council or housing association. In 1981, the average income of council tenants as a percentage of the national average was 73 per cent (Page, 1993). By 1991 it had fallen to 48 per cent. This residualisation of council housing has brought about major changes in who lives in the sector. There are now many lone parent households headed by women, many older women, and many young households living in council housing. Thus, housing issues are very much women's issues. Council tenants have a common interest as tenants but also often as women living in man-made environments and dealing with male-dominated housing authorities. Tenant participation is thus something which directly addresses women's lives.

WHAT IS TENANT PARTICIPATION ABOUT?

Recognition of the importance of tenant participation in all aspects of housing is not new. In 1959 a Central Housing Advisory Committee report, *Councils and Their Homes*, said that local authorities needed to have a better relationship with their tenants (CHAC, 1959). In 1968 the Ministry of Housing issued a circular on council housing rents which recommended that there should be more tenant consultation (Ministry of Housing and Local Government, 1968). In 1975 tenants gained the right (where their local authority agreed) to set up tenant management cooperatives. By the mid-1970s some 12 per cent of local authorities in England and Wales had formal methods for involving tenants, including participation in committees and regular discussion meetings, and altogether some 44 per cent had 'irregular discussion meetings' with tenants (Richardson, 1977).

The 1980 Housing Act included a 'Tenant's Charter' which for the first time gave clear statutory rights to local authority and housing association tenants in England and Wales to succession, security of tenure and consultation. The 1985 Housing Act consolidated these developments, with Section 104 placing a duty on housing authorities to provide information and explanations to tenants about their rights, their tenancies and the landlord's repairing obligations. Section 105 of the 1985 Act requires local authorities and housing associations to maintain appropriate arrangements for both informing tenants about proposed changes in housing management and for seeking their views. The definition of housing management includes management, maintenance, improvement, demolition and the provision of services and amenities.

By 1986/87, Cairncross *et al.* (1993) found that the proportion of local authorities with formal methods for involving tenants had risen to 44 per cent. The proportion using informal methods such as irregular discussion meetings had increased to 80 per cent.

The Housing and Planning Act 1986 amended the 1985 Act to give secure tenants the right to consultation over any proposals from the local authority to sell their homes to a private sector landlord. Although the law does not require a ballot, the Secretary of State's permission is needed for such transfers. Consent will be withheld if the Secretary of State believes that a majority of tenants do not wish the disposal to proceed.

The 1988 Housing Act gave a strong stimulus to local authorities to develop tenant participation. If tenants are dissatisfied with their council, they can opt for an alternative landlord. Tenants must be consulted about any disposals and transfers of ownership, and the Act introduced a requirement that tenants should be balloted about any proposed transfer to a new landlord. The possibility that their housing might be lost to the private sector if tenants opposed their council continuing as landlord made local authorities keen to ensure tenant satisfaction, participation and involvement in all aspects of the housing service.

Unfortunately, the 1988 Act also removed the right of all new, assured housing association tenants to be consulted over matters of housing management. However, the Housing Corporation issued a 'Tenant's Guarantee' for all housing association tenants which sought to ensure that they were given similar contractual rights to their previous statutory rights. In 1989 the Corporation issued 'performance expectations' for all housing associations and, although only guidance, this made it clear that housing associations were expected to demonstrate their accountability to tenants and to encourage greater tenant participation.

In 1991, charters for all the main public services were introduced by the *Citizen's Charter* white paper (Cm 1599). The Tenant's Charter requires every council to report performance against a set of standard indicators on an annual basis, publishing this information locally.

Although the need for, and importance of, tenant participation was recognised and stated by government over 30 years ago, it did not have much significance until recently. The Priority Estates Projects promoted tenant participation together with estate offices in the 1980s. Evaluations of these projects revealed increased tenant satisfaction; reductions in voids, rent arrears and turnover; and overall improvements in economy and efficiency (Emms, 1990). The approach was more widely introduced after 1985 when the Estate Action initiative was launched. This is aimed at improving the quality of life on rundown estates through refurbishment and partnership with the private sector. Consultation with tenants is required before Estate Action bids by local authorities receive approval from the Department of the Environment.

Today, council tenants in Britain seem to have a wide range of opportunities to be involved in their housing. They can remain as tenants of the local authority, with new rights to information and consultation; they can share the management of their estates with the council through estate management boards; and they can form tenant management cooperatives.

'Participation' has become a buzz word during the last two to three years, along with 'customers' and 'customer care'. For example, York City Council (1991), the pioneer of citizen's charters, promotes its housing service as 'putting the customer first'. Its tenants choice initiative lets tenants in modernisation schemes choose what is done to their homes, what fittings and materials are used, and which contractor does the work. The contractor works for the tenant, giving the tenant maximum control over the work being carried out to their home.

Citizen's and tenant's charters, and setting standards and measuring performance, are widely promoted as responses to the problems of over-bureaucratic council housing services. More controversially, the introduction of compulsory competitive tendering (CCT) is sometimes also presented in this light. However, there has been negligible tenant support for CCT, which does little either to increase tenant choice or to increase tenant influence. Overall, the climate in relation to tenant participation has changed dramatically in recent years. The Tenant Participation Advisory Service (TPAS), for example, is experiencing an ever-increasing demand from housing staff to help them find ways of developing effective tenant participation across their housing stock. But the call for help from tenants has not really changed, although perhaps now they are being *heard*. Tenants still request help with getting their landlord to listen to them and to carry out improvements to their homes and estates, with tenants themselves having some control over the decisions which affect them.

'Participation' is not that easy to get to grips with. None of the guidance, or requirements, have specifically defined 'tenant participation'. Tenants have learned to their cost that having a legal right to be consulted about housing management has done little to enable them to have any power over the decisions taken by their landlord. Council tenant rights in Britain are, in fact, narrowly defined in law, despite the potentially conflicting interests of landlord and tenant. The situation is essentially similar to the relationship between employer and worker. Indeed, in Sweden decisions about social rented housing take place within a framework of collective bargaining about rent levels and tenancy conditions (Blackman, 1989). Swedish tenants are organised on the scale of a large trade union and have important negotiating rights.

COLLECTIVE ACTION

One of the difficulties that has now been recognised is that of reaching agreement on what exactly is meant by tenant participation. Research has shown that it can be used to mean different things depending on who is using the term – politicians, housing officers or tenants.

TPAS and the Institute of Housing have agreed a definition in a joint guide to standards for tenant participation (Institute of Housing/TPAS, 1989: 19). This describes participation as: 'A two-way process involving sharing of information and ideas, where tenants are able to influence decisions and take part in what is happening.'

Cairncross *et al.* (1990) identify eight *processes* for participation: providing information; seeking information; listening; consultation; dialogue; joint management; choice; and control. Of these, the most common was found to be providing information. They comment:

> [M]any councils engage in providing information and occasional consultation only. Dialogue, joint management and control are rare. Tenants may welcome all these processes of participation, but can judge their experience of participation by the outcome as well as the process: whether they made a difference to the decisions taken or the service provided, as well as whether they were listened to or involved in joint decision-making.
>
> (Cairncross *et al.*, 1990: 2)

Holmes (1993) sees, in the extent to which tenants seek more influence in processes of change and improvement in their estates, an argument for *empowerment*, which she defines as people being 'enabled to act', both as individuals and collectively. This desire to participate and the ability to do so are strongly linked together. Devine (1988) explains this as participation 'feeding on itself':

> As people increasingly take control of their lives, so their ability to do so also increases. The challenge of having to take responsibility for decisions that make a difference is at the same time an opportunity for personal development. It is part of the process of becoming fully human. The feminist concept of empowering has a general relevance. To begin to feel powerful, having previously felt powerless, to win access to the resources required for effective participation and learn how to use them, is a liberating experience. Once people become active subjects, making things happen, in one aspect of their lives, they are less likely to remain passive objects, allowing things to happen to them, in other aspects.
>
> (Devine, 1988: 158–9)

Participation is especially about fairness: being equal as a person and being kept in the picture. Formal structures and processes such as committees and rules are important but secondary to sharing information, ideas and decisions. A 'culture of participation' can take years to build. It must be planned and developed over the long term. Above all, there must be some *collective* input of ideas, with contact and negotiation taking place with *representa-*

tives of tenants. This is most often achieved through tenants' associations and tenants' groups. The advantages of such collective action are summarised by Holmes (1993: 38) as:

1 There is strength in numbers.
2 You get more ideas.
3 You get a spokesperson.
4 Local authorities are geared to respond to groups.

Collective action, however, has been of most significance in struggles for *rights*. In the 1970s housing authorities were trying to consult, and involve, their tenants. Tenants' groups were forming up and down the country, and 'tenant's charters' were being produced by tenants' federations. In 1977 the National Tenants Organisation (NTO) was formed around a national campaign by tenants for stronger legal rights, and in 1978 the NTO launched its Tenant's Charter.

Much of the NTO charter's emphasis was on a change in the landlord–tenant relationship: a change in the balance of power. Tenants were asking for proper rights to security of tenure; to improved repair services (with time limits, effective enforcement and compensation for tenants); to consultation and information; and to comprehensive tenancy agreements. The charter began, 'The right to decent housing is a fundamental right. Tenants are entitled to a secure home in good condition designed for their needs at a cost they can afford in an area having proper community facilities.' Tenants have continued to be concerned with the same issues: the overall shortage of housing; the attitudes of housing authorities, central government and local councils to tenants; failures of repair and maintenance services; and the right of tenants to have a say in the decisions which affect them.

In 1991, the National Tenants Organisation and the National Tenants and Residents Federation came together to develop a new Tenant Participation Charter. This charter, launched at the first European Tenant Participation Conference in Glasgow in October 1992 states, 'There must be an equal partnership between tenants and landlords. Tenants should be treated as important customers with equal status to home owners.' It goes on to say, 'The resources necessary to house the homeless and to provide everyone with a well-maintained and managed, safe and secure home, suitable for their need at a price they can afford, are essential to fully achieving these aims.'

The tenants' agenda is clear. They want the right to have their views taken into account at all levels, they want to have the opportunity to influence the decisions which affect them, and they want the right to have *collective* involvement: to have the support, information and resources to make it possible for tenants to get together to form tenants' forums, community groups, tenants' associations and tenants' federations. Tenants want to be able to negotiate with, and to work cooperatively with, their landlord. They want this to be a *right*, not a favour graciously granted by the landlord. They also want the right to become involved in ways which suit them, with properly representative structures, and to negotiate the level to which they become involved.

All this implies a shift in the balance of power between landlord and tenants. It certainly means that landlords will not be able to dictate the terms and set the agenda and timescales from their perspective. Many local authorities and housing associations have recognised this fact.

However, in spite of the long-standing recognition of the importance of consulting and involving tenants in housing management, maintenance and design, numerous reports over the years have showed that it has not been happening. Although increasing prominence has been given to tenant participation by the Housing Corporation and the Department of the Environment, with pressure on local authorities and housing associations to demonstrate positive action, many tenants still say that they are as excluded as ever from the decisions which affect them.

Cairncross *et al.* (1990) in a national survey of 1000 tenants found that 52 per cent of respondents agreed with the statement that the council does not keep tenants informed. Over three-quarters of tenants (76 per cent) thought that they should have more say in important decisions by the council which affected them. Only 27 per cent agreed with the statement that the council involved its tenants in decision-making. So what is going wrong?

The challenge facing housing authorities is that they will have to develop a more open, equal and less paternalistic relationship with their tenants. This challenge means that ways will have to be found of producing effective information for tenants and of developing channels of communication so that opportunities are created for involving tenants at the very start of the decision-making process. It also means good relationships. Good contact will have to be developed with tenants using a variety of different methods to match different situations. Most important, it means having the commitment to be prepared to give influence to tenants throughout the decision-making process and in relation to *all* decisions. The growing influence of 'customer care' can have a positive influence in this respect because it emphasises the quality of service. In Sweden, this has been seen in terms of moving social housing companies away from operating like a manufacturing industry to operating as a service. Lindberg and Karlberg (1988) summarise this as follows:

> Traditional public housing managers had focused their attention on the bricks and mortar. At worst there was the impression that the companies saw it as their duty to protect the apartments from their tenants. It is not possible to view tenants as a nuisance and still be a successful housing manager. Tenants are strongly dependent on management and they rightly expect management to help remove the negative effects of their dependence. In other words: fundamental changes in attitudes were essential at all levels in the public housing companies.
>
> (Lindberg and Karlberg, 1988: 87)

In Britain, this is an even greater problem because of a tendency to see tenants as recipients of welfare rather than as participants in their own housing. In seeking to involve tenants, however, there have been two main problems up until now. Either the landlord retained control and power, while trying to involve tenants on terms which suited the needs of the landlord more than the needs of the tenants, or the enthusiasm of the landlord drove them forward at a pace that was too fast for the tenants, and the tenants were not ready for the structures created for participation by their landlord. Landlords failed to appreciate the need to help build up the confidence of the tenants. They also failed to appreciate the necessary evolutionary process of participation.

Holmes (1993) stresses that tenants need to acquire gradually the skills and competencies to participate. She identifies six main factors which can put local people off getting involved:

1 Feeling that they will not be effective.
2 Experiencing failure.
3 Low self-esteem and lack of confidence.
4 Not wanting to fuss.

5 Not liking the other people involved.
6 Domination by 'experts' and 'professionals'.

It follows from this that people will get involved if they feel they will be effective, have experienced success, see peers involved, feel some personal benefit or face something really bad! But even unjust or harmful action by the council will not necessarily mean that tenants respond with collective action. Cairncross *et al.* (1993) found that most tenants in a national survey said that they would deal with such problems by contacting their councillor, MP or a council officer, or by signing a petition. Tenant participation for most tenants is about supplementing and strengthening traditional democratic mechanisms with more sensitive ways of giving people a voice when decisions are being taken which affect them.

The decentralisation of housing services in many local authorities has often enhanced contact between tenants and council landlords. However, decentralisation alone will not enable participation, especially among people who are most disadvantaged by poverty or discrimination. It is often the case, for example, that there is a dearth of local leadership or organisation on housing estates. In other cases, there may be a plethora of groups and too many voices, with the problem of knowing who represents whom. For these reasons, community development is an essential strategy for realising tenant participation and bringing diverse groups together, in particular extending the support of resources and community workers to those areas and groups where discrimination, disadvantage and feelings of powerlessness are greatest (Association of Metropolitan Authorities, 1993).

Another key problem for landlords concerns the requirements laid upon them by central government. Central government, via the Department of the Environment and the Housing Corporation, expects housing authorities to be giving information to their tenants, and to be developing tenant participation. In order to encourage housing authorities to develop tenant participation, the allocation of resources to them is becoming more dependent upon the level to which they have succeeded in developing that participation. Central government is dictating clear terms and conditions to housing authorities in relation to the provision of resources and the development of tenant participation. Unfortunately, these terms and conditions bear little resemblance to what tenants themselves want and have been seeking for over 20 years. The resources for local authorities and housing associations to carry out repairs and improvements, and to build new homes at affordable rents, have decreased not increased.

Rights for tenants to be consulted about rents and financial matters were specifically excluded from the government's Tenant's Charter. Its 'right to repair' did not provide a quick, easy or effective way for tenants to get an improved repair service. Not only was this tortuously complicated with a restricted cost limit, it also involved tenants having to pay for the repair before they could claim money back. It did not give any rights to tenants to negotiate a better repair system from their landlord, which was what they wanted!

Local authorities are required to produce reports to tenants about their performance, but with information which tenants have not asked for. They are not required to produce the local information, or information about policies, which tenants *have* asked for. Central government is making resources available if tenants want to set up tenant management boards. But unfortunately resources are not available to help tenants develop tenants' groups, local forums or other less formal and bureaucratic structures. Resources are also not available for continuing training for staff and tenants, which everyone has acknowledged they need.

Because the resources might be available if housing authorities do what central government wants, both tenants and landlords fall into the trap of developing structures which are more likely to exclude the majority of tenants. All available evidence suggests that only a small percentage of tenants want to take on all the responsibilities of running their housing themselves. Only the most confident, experienced and articulate tenants are likely to survive the intensive development of formal management bodies.

A problem for tenants themselves is knowing how to go about getting people involved and knowing how their landlord actually takes decisions. Many tenants lack the confidence to become involved and many also do not believe that they themselves can make any difference. Tenants also have unrealistic expectations about the level of continuing involvement that they should expect from other tenants in the area. Often inadvertently they take their landlord's expectations as their guide.

In order to find a way to become involved in the decision-making of their landlord, tenants can find that they are being forced to mirror the committee structures and procedures of their landlord, even though they know that these very structures help to keep out effective input by tenants.

BREAKING DOWN THE BARRIERS

Tenants' organisations have little official recognition from central government. There is no formal structure to ensure that tenants' organisations are consulted by central government, and no way in which tenants can have any formal influence over any policy development which might affect them. Proposals for legislation and guidance for housing authorities which will have a direct impact on tenants, and which are often supposed to benefit tenants (the 'Right to Repair', 'Tenant's Choice', the 'Tenant's Guarantee', the Tenant's Charter, the 'Right to Manage', the Compulsory Competitive Tendering of housing management services, and so on), are not developed in liaison with them. Much of the recent legislation which the government has said is to enhance tenants' rights has been bitterly opposed by tenants. When consultation papers are produced by the Department of the Environment or the Housing Corporation there is no automatic right for tenants' groups or the national bodies of tenants – the National Tenants' Organisation and the National Tenants' and Residents' Federation – to be asked for their views.

Resources are being made available to promote and develop tenant participation but these resources do not go to the tenants themselves who might be struggling to develop tenants' groups. The resources go to 'agencies' which are supposed to encourage the development of estate management boards and tenant management cooperatives.

Tenants have called for help, support, recognition and resources for tenants' *groups*. They have called for *collective* rights. They have called for formal recognition at both local and central government levels so that they can be consulted about, and involved in, the development of policy and practice in relation to their housing.

Legislation has focused on *individual* rights and has failed to give any collective rights to tenants. Even worse than this, the Local Government and Housing Act 1989 removed the right of tenants to play a full part in their local housing committees. Many tenants' groups had their own elected tenant representatives playing a full part in local sub-committees, and even in the main housing committee. The 1989 Act removed their right to have any sort of vote or be full members of such committees.

Although most of the legislation was the opposite of what tenants wanted, it has had some positive effects. Much of the legislation was perceived by tenants to be a direct threat to their local authorities as landlords. Although tenants had been demanding a better deal, they had not been demanding a change of landlord. Tenants had been asking for a better relationship with their local authority landlord, with more power for the tenants.

Local authorities realised that if they did not have a better relationship with their tenants they might lose their housing services. This has resulted in many local authorities making serious efforts to get together with their tenants and to develop tenants' groups. Housing associations also realised that the spotlight had turned on them, showing a poor track record in relation to tenant participation, and so they in turn are now actively trying to encourage the development of tenants' groups. Unfortunately, legislation hinders rather than helps in that local authorities cannot decide to make tenants full voting members of committees. Nevertheless, both local authorities and housing associations are now looking at ways in which they can develop accountability to tenants and ensure that tenants do have influence over the decisions which affect them.

In spite of the mistakes that have been made in the past, there are now many examples of good practice to be found around the country. This is developing in towns and villages as well as in cities and urban areas. There is a lot that can be learned from both the mistakes and the successes. What is perhaps most interesting is the enthusiasm of tenants for getting together with other tenants to share experiences and ideas with the aim of working positively with housing staff, councillors and committee members, in spite of all the constraints and decreasing resources.

There are a number of common issues which seem to generate tenants' groups and interest in getting together to take action. These are as follows:

1 Lack of affordable housing in the area.
2 Repair problems.
3 Broken lifts in flats.
4 Dirty, broken, unsafe paths.
5 Lack of play space.

6 Security problems.
7 Traffic problems.
8 Need for improvements or modernisation.
9 Feelings of isolation and the need for a 'community'.

It is often women who get together first. Women are most likely to be affected by these problems: *younger women* (often lone parents) at home with small children and *older women*, alone, often after the death of a partner. Both groups frequently have only some form of state benefit as their income. They are often faced with few opportunities for social activities, in both inner city and rural areas. It is not surprising, therefore, that a high proportion of tenants' groups are started by women.

Particular problems face women who want to become involved in tenants' groups. These are:

1 Lack of availability of childminders and childcare.
2 Unsuitable times and venues for meetings (poor access and facilities, poorly lit streets and areas and meetings held when it is late and dark).
3 Lack of confidence.
4 Lack of experience of formal meetings.
5 Intimidating and bureaucratic procedures.
6 The attitudes of men in authority.
7 Men moving into the 'power-base'.
8 Becoming part of a bureaucratic structure.

Women often begin by getting together and sharing out tasks among themselves, without having formally elected 'officers'. Tenants' groups often start informally, and often manage very well at the start without formal rules, constitutions and standing orders. However, local authority and housing association landlords operate formally, especially in relation to decision-making. They are quite correctly concerned about making sure that any tenant participation is on as representative a basis as possible, and they usually only have knowledge and experience of achieving anything through a structure controlled by rules, standing orders, constitutions and formal frameworks. The difficulty, therefore, is to find a way of ensuring that some form of representative and accountable structure develops, without too tight a formal framework that defeats the continuing input of the women who have started the group.

In a tenants' group – especially groups that become powerful via campaigns about housing action trusts, transfers, and so on, and those at national level – a man is more often seen as the spokesperson and especially as the chair. Many of the groups that have acquired a role within their landlord's decision-making structures have men as their spokespersons, and on the committees. Many of these groups also say that they are having problems in involving younger tenants, younger women with children and Black people. If men have more experience of organising formally and of meeting with and speaking to those 'in charge' via their work experience, and often trade union experience, it is hardly surprising that they take over tenants' groups.

WHAT ARE THE LESSONS?

The benefits of tenants' associations and of the involvement of tenants' groups in housing management have been well researched and documented. There are some benefits that are particularly important. These are the benefits which women themselves say they have gained: mutual support; sharing ideas and learning from each other; gaining confidence; recognition of one's own skills and the value of one's own experience; and the development of support networks. As one tenant explains:

> By belonging to the tenants' association, I've seen a lot of things I never knew existed and it's made me much more aware of other people's problems. I think it's made me a better person. I feel now that I could go out and cope with a lot of things I couldn't have done before. For instance, I never could have stood up and spoken in front of people. I would've been too nervous. And now I've joined the school governors and I never would have done that, because I wouldn't have thought I was good enough. It's made me feel better about myself and made me realise it's only everyday, ordinary people who do these things.
>
> (Institute of Housing/TPAS, 1989: 88)

Women have always played a key role in tenants' groups but this has not necessarily meant that they are able to be in, and stay in, positions of power and control. If the problems and disadvantages outlined in this chapter are to be overcome, there are some important things which must be done. These are:

1 Support, resources and training – not tied in to forms of participation prescribed by the landlord or central government.
2 Flexibility of approach.
3 New structures and forms of organisation – especially recognition of the need to allow structures to evolve at a pace with which tenants can cope.
4 More women in key jobs.
5 Training for housing staff, councillors, committee members and tenants on a continuing basis.
6 Formal recognition of tenants' groups both locally and nationally.

Above all, what is needed is recognition of the fact that participation is about working cooperatively together. This means that tenants themselves must be able to influence the timescale and agenda for participation. It also means that a more equal relationship, with real equality of opportunity for all tenants to take part, must be developed. The policies for tenant participation must come from what tenants are really asking for, not what central government (or housing authorities) think they should have. These changes are needed if women are to be able to play a full part as 'customers' in our new world of tenant participation.

It is not clear what the new structures and relationships will be. But they must be developed in order to break out of the bureaucratic strait-jacket that has not only hindered the development of real tenant participation but helped to keep women in a less equal position. Participation cannot work if the partner with the power says, 'you can only participate on my terms, in the way that I say, in the things that suit me'.

REFERENCES

Association of Metropolitan Authorities (1993) *Local Authorities and Community Development: A Strategic Opportunity for the 1990s*, London: AMA.

Blackman, T. (ed.) (1989) *Tenant Influence in Public Housing: Case Studies of Sweden and Northern Ireland*, Coleraine: University of Ulster.

Bondi, L. and Peake, L. (1988) 'Gender and the city: urban politics revisited', in J. Little, L. Peake and P. Richardson (eds) *Women in Cities*, Basingstoke: Macmillan.

Cairncross, L., Clapham, D. and Goodlad, R. (1990) 'Tenant participation in council housing', *Housing Research Findings No. 8*, York: Joseph Rowntree Memorial Trust.

Cairncross, L., Clapham, D. and Goodlad, R. (1993) 'The social bases of tenant organisation', *Housing Studies*, Vol. 8, No. 3: 179–93.
CHAC (Central Housing Advisory Committee) (1959) *Councils and Their Homes*, London: HMSO.
Devine, P. (1988) *Democracy and Economic Planning*, Cambridge: Polity Press.
Emms, P. (1990) *Social Housing: A European Dilemma?*, Bristol: School for Advanced Urban Studies.
Holmes, A. (1993) *Limbering Up: Community Empowerment on Peripheral Estates*, Middlesbrough: Radical Improvements for Peripheral Estates (RIPE).
Institute of Housing/TPAS (1989) *Tenant Participation in Housing Management*, Coventry: Institute of Housing/TPAS.
Lindberg, G. and Karlberg, B. (1988) 'Decentralisation in the public housing sector in Sweden', *Scandinavian Housing & Planning Research*, 5: 85–99.
Ministry of Housing and Local Government (1968) *Increase in Rents of Local Authority Housing: Prices and Incomes Act, 1968*, Circular 37/68.
Page, D. (1993) *Building for Communities: A Study of New Housing Association Estates*, York: Joseph Rowntree Foundation.
Richardson, A. (1977) *Tenant Participation in Council Housing Management*, London: Department of the Environment.
York City Council (1991) *Putting the Customer First: York's Approach to Customer Service in Housing*, York: York City Council Housing Services.

Chapter 5

Older women and housing – prospects for the 1990s

Roger Sykes

INTRODUCTION

Elderly people have over many years become increasingly marginalised. This is seen in their housing circumstances, incomes and opportunities. The 1980s saw a real decline in the housing and living standards for large numbers of elderly people. The reality was not affected by government attempts to counter it by promoting the notion of the 'Woopie' (well-off older person).

In the 1980s this term created new images of elderly people with greater spending power, greater housing choice and greater opportunity throughout their retirement. This view was expressed by the former Secretary of State for Social Security, John Moore: 'It is simply no longer true that being a pensioner tends to mean being badly off. . . . The modern pensioner has a great deal to contribute and a great deal to be envied.'

This 'rose-tinted' image masks the diversity of circumstances in which elderly people find themselves. For the vast majority of elderly people real housing choice and income security will remain elusive aspirations. Many elderly people find themselves at the margins of society, dependent on state benefits and in many instances severely disadvantaged.

Elderly women as the largest group within the population of older people find themselves in the most vulnerable position. Lower incomes, poorer housing conditions, longer life expectancy and higher levels of disability are major problems facing older women. These will combine to provide challenges for social policy makers and providers of services in the 1990s. Providers have not necessarily responded to their full potential in seeing older women as a group with specific needs.

The discussion below provides factual evidence to illustrate the circumstances of elderly women. From this it seeks to draw out some policy priorities for the coming decade.

THE DEMOGRAPHIC BACKDROP

Women make up around two-thirds of the elderly population. Data from the 1991 Census (Table 5.1) show there are over 10 million elderly people in Great Britain representing 18.7 per cent of the population.

Table 5.1 Pensioner population profiles 1991

Total population	Total pensioner population	Pensionable age to 74	75–84	85+
54,888,844	10,264,213	6,421,994	3,018,886	823,333

Source: OPCS Census Monitor: Great Britain, 1991

Table 5.2 Pensioner profiles by age 1991

Pensionable age to 74	75–84	85+	75 and over	
			Males	Females
11.7%	5.5%	1.5%	2.4%	4.6%

Source: OPCS Census Monitor: Great Britain, 1991

While the proportion of the 60/65–74 age group has actually fallen over the past decade, significant growth occurred in the number of people over 75, with even more pronounced growth in the over-85s. Table 5.2 shows the percentage proportions of older people and the larger number of older women among the over 75s.

This contrast in the proportion of elderly women to men has been studied in recent research examining the processes of ageing and their impact on welfare services (Johnson and Falkingham, 1992). This work has investigated historical changes in gender differences over the past century and the major increase in the

Table 5.3   Female to male population ratios

Year	Female : male ratio 60/65+	Female : male ratio 85+
1881	1.97	1.64
1901	2.05	1.74
1921	2.04	2.07
1941	2.03	2.20
1961	2.26	2.32
1981	2.02	3.26

Source: Johnson and Falkingham, 1992

numbers of single elderly women. The lower mortality rates among women compared to men in middle and later life is the primary cause of this pattern. Table 5.3 illustrates these changes. The high ratio of elderly women among the over 85s is particularly marked.

HOUSEHOLD STRUCTURE

There are a number of key demographic and family structure changes among the older population and their families which are having a growing impact on the future provision of housing.

There has been a noticeable increase in the number of elderly households. In particular there has been a growing trend to more elderly people living alone, and the vast majority of these are women. In 1985 there were 5.3 million elderly households and this is set to increase to 5.8 million by 2001.

The major growth area over the coming decade will be single person households – there will be 620,000 more single person households by 2001. The vast majority of these will be widowed women.

The proportion of older people living alone has increased substantially since the turn of the century. In 1901 around 10 per cent of older people lived alone. Table 5.4 shows differences in people living alone by gender and age and reinforces the picture of differences in older age between men and women; 61 per cent of women over 75 live alone.

Table 5.4   People living alone by age and sex (% figures)

	Men	Women	Total
16–24	4	4	4
25–44	8	5	6
45–64	10	13	11
65–74	17	36	27
75 or over	29	61	50

Source: General Household Survey,1988 (HMSO, 1990)

HOUSING TENURE

The housing tenure of the population as a whole has been transformed over the past 50 years. There has been a steady decline in
the private rented sector, a growth and then decline in public
sector housing, and a steady and continuing growth in owner
occupation. The 1980s saw a massive increase spurred on by
government policy proclaiming the virtues of a 'property-owning
democracy' and accelerated by the large-scale sell-off of local
authority houses under 'Right to Buy'. Table 5.5 shows the
changes in tenure patterns over recent years.

Table 5.5   Housing tenure in Great Britain

Year	Owner occupier		Public sector rented		Private sector/ housing association rented		Total dwell-ings
	000s	*%*	*000s*	*%*	*000s*	*%*	
1961	6885	42	4201	26	5187	32	16,273
1971	9610	51	5811	31	3578	18	19,000
1981	11,936	57	6387	30	2755	13	21,078
1991	15,517	67	5049	22	2418	11	22,983

Source: Department of the Environment, *Annual Report* (HMSO, 1993)

At present around half the older population is in rented accommodation and half in owner-occupied property. The proportion
of older owner occupiers will increase in subsequent decades.

The proportion of older people living in rented accommodation is greater for the over-80s, and these are older women.

HOUSING CIRCUMSTANCES

Elderly people find themselves disproportionately in poorer housing than other age groups. This pattern is common across different housing sectors. The 1986 *English House Condition Survey* (DOE, 1988) presented an initially optimistic picture of improving housing conditions. However, on closer examination elderly people still live in some of the poorest housing conditions:

- Elderly people living alone, mainly women, are more likely to occupy housing in poor condition than elderly couples.
- Poor housing conditions are strongly related to length of residence. Those people who have lived in the same property for over 20 years, mainly elderly people, are more likely to be living in poor housing conditions.
- Housing conditions vary considerably with tenure. Some 42 per cent of private rented tenant properties were in poor condition; 13 per cent of owner-occupied dwellings; 11 per cent of those owned by local authorities.

While there has been some improvement in reducing the number of elderly households which lack basic amenities, the overriding problem of large numbers of elderly people in poor housing remains. When we examine the over-75 population, there are 112,000 houses lacking basic amenities, 116,000 houses unfit for human habitation by the standards of the local authority, and 275,000 houses in poor repair. For the whole elderly population there are over 1.3 million elderly people living in poor housing conditions in urgent need of assistance.

When this information is put alongside the high proportion of women in the over-75 population we are left with a picture showing older women, mainly single, being in considerable housing need.

THE FINANCIAL CIRCUMSTANCES OF OLDER PEOPLE

There has been much discussion about the wealth and income of elderly people. Much of this government-inspired debate has focused on the new prosperity of elderly people. However, the reality is very different.

Recent work has produced a typology of the income characteristics of elderly people. Two separate pieces of research (Bull and Poole, 1989; Bosanquet et al., 1989) have clearly identified three broad groups of elderly people. While generalizations are always fraught with some problems, the typology does produce an important framework to look at the housing options available to elderly people.

The first group identified were the well-off representing around 20 per cent of the elderly population. These people have adequate resources to cover increases in expenditure caused by declining health and reduced mobility, changing housing needs or the need to buy in care services. These people tend to be those who have been in well-paid employment during their working lives. They will have enjoyed higher salaries, have enhanced pensions, money in investments, high value properties and considerable savings. This group will not need to call on the support of the welfare state.

The second group represents around 40 per cent of the elderly population. This is the group studied in more detail in the 'Not rich, not poor' research (Bull and Poole, 1989). These are elderly people on middle incomes. The vast majority of this group are owner occupiers who have paid off their mortgages. The group is characterised by many who are asset-rich, but income-poor households. Many have limited incomes or savings which are just above benefit thresholds. This, in effect, disenfranchises many elderly people from state support. While these households may be materially better off, their funds are often insufficient to allow them to exercise real choice in taking advantage of different housing options. The group is characterised by large numbers of single elderly women living on their own, often with a small occupational pension.

The third group are the substantial number of elderly people who remain poor and extremely disadvantaged. They will be on low incomes, and highly dependent on state benefits to raise income levels. This group tends to be in the rented sector, either local authority or privately rented, and so fails to have the benefit of equity within an owner-occupied property. This group is likely to be the one most at risk from poverty and poor housing conditions.

From this typology, it is clear that around 80 per cent of elderly people have limited resources which will restrict their choice in retirement as their circumstances change through ageing, increased frailty and dependency. See Table 5.6.

Table 5.6 Typology of not rich, not poor

	Well-off 20%	Not rich, not poor 40%	Poor 40%
Income	High	Middle to low	Low
Savings/ investments	Substantial	Limited	Limited
Tenure	Owner/occupiers	Mainly owner/ occupiers	Local authority or private tenants
State benefits	None	Partially dependent	Highly dependent
Occupational pensions	High levels	Small amounts	Low

Source: Bull and Poole, 1989

Income characteristics

The income of elderly people is drawn from four main areas: the retirement pension and other associated state benefits; occupational pensions; savings and investments; and wages.

There are evident and clear income inequalities among the older population. Table 5.7 illustrates the main characteristics of those most likely to experience poverty in old age.

It is also clear that income will vary considerably throughout an older person's retirement:

• income tends to decline with the death of a spouse, while expenditure is seldom significantly reduced;
• expenditure increases with age, yet the 'old-old' have the lowest real incomes;
• older women tend to find themselves with lower income levels throughout retirement.

Table 5.7   Income inequalities in older age

- Over-75s
- Older women
- Those living alone
- Those who do not own their own home
- Ethnic minorities

Retirement generally means a reduction in real terms in income levels and it is likely that income will reduce with increasing age.

Income distribution

Since the late 1970s the real value of pensions and therefore older people's incomes has been steadily eroded. Before 1980 the increase in state pensions was linked to average increases in earnings. In 1980 pension increases were linked to the retail price index. This has meant that the real value of the pension has been reduced throughout the 1980s. As a result elderly people find themselves disadvantaged compared to those in employment. It means that their disposable income has been reduced and many have been left in considerable poverty. See Table 5.8.

Table 5.8   Changes in retirement pensions

	Single £ per week	£ per annum	Married £ per week	£ per annum
1982	£32.85	£1708.20	£52.55	£2732.60
1984	£35.80	£1861.60	£57.30	£2979.60
1986	£38.70	£2012.40	£61.95	£3221.40
1988	£41.15	£2139.80	£65.90	£3426.80
1990	£46.90	£2438.80	£75.10	£3905.20
1992	£54.15	£2815.80	£86.70	£4508.40

Data from the most recent Family Expenditure Survey show that many people remain on low incomes. While a growing proportion of the older population are experiencing greater incomes, when income levels are compared for different household types the differences become more evident.

Table 5.9 Gross weekly income of all retired people

Under £60	17%
£60 < £80	13%
£80 < £100	13%
£100 < £125	11%
£125 < £150	9%
£150 < £175	6%
£175 < £225	8%
£225 < £325	10%
£325 and over	13%

Source: *Family Spending: A Report on the 1990
Family Expenditure Survey* (HMSO, 1991)

Table 5.9 shows that around a third of elderly households have incomes of less than £100 a week and they are likely to be highly dependent on state benefits.

When single elderly person households are compared to the total population inequalities are noticeable.

Table 5.10 Weekly household income levels at different ages

	Under £100 %	£100–225 %	£225+ %
15–24	22	31	47
25–39	10	16	74
40–49	9	11	80
50–59	11	19	70
60–64	22	28	50
65–69	29	36	35
70–74	39	39	22
75+	50	33	17

Source: *Family Spending: A Report on the 1990 Family
Expenditure Survey* (HMSO, 1991)

Table 5.10 illustrates the substantial income differences between those in employment and those in retirement. There is a severe fall in income on retirement and noticeable growing poverty with increasing age. This trend is intensified for older women.

Table 5.11   Distribution of weekly household income by retired household type

	Under £100 %	£100–225 %	£225+ %
One adult (mainly dependent on state pension)	96	4	–
One adult (other)	31	49	20
One man, one woman (mainly dependent on state pension)	43	57	–
One man, one woman (other)	2	53	45
All retired	43	35	22

Source: *Family Spending: A Report on the 1990 Family Expenditure Survey* (HMSO, 1991)
Note: Pensioner household income 'mainly dependent on state pensions' is where 75 per cent of income is from state pension and housing and other benefits.

A significant proportion (around 60 per cent) of elderly people, particularly women, are highly dependent on the state pension for 80 per cent of their income. For 20 per cent it is their sole source of income alongside state benefits.

There are significant differences between household types. The distribution of income according to household type is illustrated in Table 5.11.

There are further significant differences in the sources of income among older people. In many instances these differences reflect the housing, employment and income characteristics of older people in their working lives. It is clear that single elderly people are more likely to have lower incomes and in many instances this is because of the pattern of their working lives and the lack of additional income through occupational pensions. See Table 5.12.

Occupational pensions

Much of the focus on the supposed growth in affluence among elderly people has concentrated on the impact of increasing income levels of occupational pensions. To a lesser extent personal pensions will also have an impact. However, their effect in raising income levels of older people has been greatly overstated.

Recent research into retirement trends has highlighted income inequalities experienced by many elderly women. There has been a trend in recent decades towards earlier retirement. This has been

Table 5.12   Source of income of different retired households

	Social security benefits %	Savings %	Annuities/ occupational pensions %	Other %
One adult (mainly dependent on state pension)	82	3	4	11
One adult (other)	30	29	27	14
One man, one woman (mainly dependent on state pension)	83	3	5	9
One man, one woman (other)	32	20	30	18
All retired	41	17	20	22

Source: *Family Spending: A Report of the 1990 Family Expenditure Survey* (HMSO, 1991)

Notes: 1   Other includes income from wages, imputed income.
2   Pensioner household income 'mainly dependent on state pension' is where 75 per cent of income is from state pension and housing and other benefits.

more prominent among men: 60 per cent of men retire before they reach 65, whereas only 33 per cent of women retire before 60. Women are also more likely to retire after the retirement age than men.

The occupational pensions received by men and women show considerable variation. This is because of a number of employment trends. Women, on the whole, tend to be in lower paid employment than their male counterparts. They are also likely to have more breaks from work when raising a family and return often only on a part-time basis. While there is some evidence that these trends are changing for today's female workforce, this was not the case for the generation who are now retired.

These trends have meant that the occupational pension levels of women are much lower than their male counterparts, or in many instances the receipt of an occupational pension is lost on the death of the male spouse, leaving the woman with further reduced income. Table 5.13 illustrates these marked income variations in occupational pension payments. There is considerable variation in those people who receive an occupational pension.

Table 5.13   Mean weekly income of elderly men and women with and without occupational pensions

Man with occupational pension	Woman with occupational pension
£126	£78

Man without occupational pension	Woman without occupational pension
£72	£40

Source: Bone, 1992

The survey (Bone, 1992) showed that two-thirds of retired men received an occupational pension, whereas only a quarter of women received one. The average amount drawn also varied considerably – men received £61 a week, whereas women received only £30 a week.

The inequalities in occupational pensions are not only noticeable between men and women, but also according to employment type. While both male and female manual workers are less likely to have an occupational pension, or for it to be at a low level, female manual workers are even less likely than their male counterparts to have this as a source of income. This is again largely owing to shorter contribution records, thus involving lower occupational pension entitlements. There is growing evidence that many women, who worked part time, decided not to join an occupational pension scheme as this would have represented a further reduction in income in part-time employment. Those people who received the highest incomes in their working lives will generally receive the highest retirement incomes. Women will have lower income in retirement as a result of lower paid, broken employment during their working lives. See Table 5.14.

The 1980s saw the promotion of personal pension plans and around 3.5 million policies were sold. The vast majority were sold to people in their 20s and 30s. Hardly any policies were sold to men over 50 or women over 45. It is likely that personal pension plans will only begin to make an impact on retirement incomes for those people retiring from 2020 onwards.

Table 5.14   Average earnings of full-time
employees

	Average weekly earnings (gross)
Males	
All males	£241.60
Manual males	£196.60
Non-manual males	£290.90
Females	
All females	£161.60
Manual females	£121.90
Non-manual females	£172.60

Source: New Earnings Survey 1988 (HMSO,
1990)

Savings and investments

Savings levels among the current generation of older people are
typically low. Table 5.15 shows the savings levels of both men and
women based on an OPCS survey on retirement and retirement
plans. This shows that women are likely to have lower savings
levels than men – 71 per cent of women have savings under £6000
compared to 58 per cent of men.

Table 5.15   Value of savings and investments of
retired men and women

	Men %	Women %
None	22	25
Under £3000	29	36
£3000–£6000	8	10
£6000–£8000	5	4
£8000–£10,000	5	4
£10,000–£20,000	10	7
£20,000–£30,000	4	3
Over £30,000	12	6
Missing	6	6
	100	100

Source: Retirement and Retirement Plans (OPCS, 1992)

DISABILITY IN OLD AGE

Elderly people are likely to experience higher levels of disability than the rest of the population and recent research has confirmed the extent of these disability levels. Two recent reports by the OPCS on the prevalence of disability identified large numbers of older people with some form of disability. The survey focused on disability restricting the ability of a person to perform everyday tasks. It showed that even minor disabilities can make existing housing unmanageable or potentially dangerous. The survey revealed that older women in particular were more likely to suffer greater disability levels.

The survey identified over 6 million people in Britain as having some form of disability. Of these, 4.2 million were older people. This represents around 40 per cent of the elderly population. Of the 4.2 million elderly people identified, 90 per cent lived in general housing.

The research demonstrated that disability increases dramatically with age. Seventy-five per cent of older people over 80 were likely to have some form of disability. The vast majority of these people will be older women living alone.

Elderly people are also most likely to have the severest disabilities of any age group. Sixty-four per cent of those people in the severest categories of disability were aged 70 and over.

Older women are more likely to suffer disabilities than older men. They also tend to be those who are on the lowest incomes and most dependent on state support. Single elderly women with a disability are identified as the most vulnerable and disadvantaged disability group. This group has experienced higher living costs as a result of their disability, while having low incomes which cannot be increased.

WOMEN AS CARERS AND THOSE CARED FOR

One of the major social policy areas for the 1990s is the position of carers in society. The past 30 years have seen fundamental changes within family structures, and also changed expectations of women and of their role. Family sizes are now smaller; women are having children later in life, and women are wanting greater independence and careers. At the same time, there is an increasing number of elderly people with greater levels of disability and frailty. It has typically been women who fulfil the caring role, but there is some evidence that this trend is changing. The most comprehensive recent research was conducted by the National Carers Survey Research Team. Their report produced the following key findings:

- 73 per cent of older people felt that they would need care in the future, but did not want to burden their family with the caring responsibility;
- the majority of older people would prefer to remain at home with care provided in their own homes;
- 40 per cent of carers said that they had to give up work to care full time as they could not combine caring and employment.

While the family still remains the main source of support and care for older people, this trend is being weakened.

There are larger proportions of women working than before and this is likely to continue throughout the 1990s, despite the current recession. Traditionally, the role of caring for older people has fallen on middle-aged or elderly children, usually daughters. The numbers in this pool of unpaid care, mainly women between 45 and 60, are declining. This group is also staying longer in part-time and full-time employment, and so is unable to provide the same levels of informal care.

There has also been a noticeable change in the geographical mobility of people around the country. This has meant that traditional, close-knit, local family networks have been eroded.

With the possible decline in the numbers of informal carers there will be more pressure on statutory services to deliver care. However, with the decline in the number of school leavers, there is a further potential problem of a scarcity of young people coming into the caring professions (nursing, etc.).

HOUSING OPTIONS

Our account so far has presented the housing, care and income characteristics of the older population with particular focus on the circumstances of older women. The account has tried to demonstrate the range of conditions in which older women find themselves, and it is these conditions which will shape their housing options in their retirement.

One must remember that the vast majority of older people live in general housing and will continue to do so throughout their retirement. Around 90 per cent of older people live in general housing with the balance being in sheltered housing, residential care or nursing homes.

It is also important to recognise that most older people choose to remain in their own homes where, in many instances, they have lived for a considerable time. However, moving in old age is something that many older people face and the options open to them depend very much on their existing housing tenure; their income and savings; and the level of family support.

Table 5.16 attempts to present the options open to older people who move during retirement. What is most clear is that alternative housing options are strongly determined by tenure.

Table 5.16   Housing options in old age

Existing housing	Option for future
Local authority tenant	– Sheltered housing to rent (local authority; housing association) – Other, non-specialist rented accommodation
Private sector tenant	– Sheltered housing to rent (local authority; housing association) – Other non-specialist rented accommodation
Owner occupier	– Sheltered housing to rent – Smaller owner-occupied property – Retirement housing to buy – Staying put assistance

Source: Rolfe, S. *et al.* (1993) *Available Options*, Oxford: Anchor Housing Trust

Older people who are owner occupiers have, in theory, a greater choice of housing options, although this will depend on the value of their property, its geographical location in the United Kingdom, and the supply of affordable options in their price range. A number of options are available. The owner may 'trade down' to a smaller property, thereby releasing surplus cash from the sale of the larger house. The move may be to either non-specialist housing or specialist sheltered housing to rent or purchase.

Owner occupiers also have the option of adapting or improving their property through the growing number of Staying Put and Care & Repair projects across the country. These projects aim to help older home owners with repairs, improvements and adaptations to their properties. They work closely with the client to secure funding and ensure supervision of building work. Research (Mackintosh and Leather, 1992) has demonstrated the lasting effect of this type of work; it enables elderly people to remain in their own homes for longer periods and thus delays the move into sheltered housing or residential care.

The 1980s saw a massive expansion in the private sheltered housing market. Private sector builders sought a new market area and quickly found a niche for retirement housing which expanded rapidly through developers such as McCarthy and Stone and Anglia Secure Homes. This was followed by a second wave of developments by many of the mainstream builders such as Wimpey and Bovis. By the end of the 1980s there were estimated to be around 90,000 private retirement houses.

The industry broadly followed the design features adopted by the local authority and housing association sector. However, by the late 1980s, with the housing slump and subsequent recession, developers were finding it increasingly difficult to sell properties. Research at the University of Surrey has shown that the average age of purchasers was 76 and that 80 per cent of purchasers were women. Purchasers were almost exclusively women who had traded down to smaller accommodation, in many instances following the death of the husband.

A number of concerns have been voiced about this sector of housing for older people, involving over-priced properties and rapidly rising service charges. Furthermore, it has been seen that for many people the trade-down option to retirement housing was not a real one.

Older people in the rented sector are clearly, by and large, limited in their housing options to moves within the rented sector – either privately, through the local authority or housing association, or to specialist rented sheltered housing.

Two things must be realised about housing options and moving in old age. First, for many older people, whether rich or poor, whether owner occupiers or tenants, moving house is an attempt to improve the quality of their lives, and the quality and appropriateness of their housing.

Second, moving in old age is an important decision. For many older people a move will effectively reduce future house move options. For example, selling a large property to move to a retirement flat may not provide the financial trade-down sufficient to sustain service charges in the longer term and the cost of future care.

What is undoubtedly true for large numbers of older people is that real choice of housing is, in many instances, limited by the existing tenure of the property, the value of the property (if owned), geographical location, and the supply of alternative affordable housing in the vicinity.

Now we look at some of the services being offered to older women by Anchor.

DELIVERING SERVICES TO OLDER WOMEN: THE EXPERIENCE OF ANCHOR

Anchor is the largest provider of housing and care for older people in England. It currently houses more than 24,000 tenants in over 21,000 units of sheltered housing. As well as sheltered housing, Anchor also provides houses for over 2000 people in Housing with Care for frailer older people and housing advice to home owners through its 34 Staying Put projects.

Anchor aims to provide a range of housing services to older people according to their needs. Older women form the main client group which the organisation helps, but there is some variation across the services.

Sheltered housing is the main activity and Anchor has more than 600 sheltered housing developments. Typical schemes will have around 30 units, with communal facilities, emergency alarm system and resident warden. About half the stock are bedsits, with the remainder being mainly one-bedroom accommodation. There are also some bungalows and two-bedroom flats. This housing offers accommodation appropriate to the needs of vulnerable elderly people whose existing circumstances have become unsuitable.

Around 70 per cent of the lettings to new tenants in 1992–3 were to older women. This trend is reflected in the overall tenant population where 74 per cent of tenants are women. The majority of all current tenants (59 per cent) are single elderly women.

There has been a trend within Anchor's sheltered housing towards older tenants which was in part caused by changes in the tenant selection policy in 1988. There is a move away from the 'balanced' approach to sheltered housing, whereby frailer tenants were looked after by more active tenants, towards a more needs-based allocations procedure.

Since 1982 the average age of all Anchor's tenants has increased from 75.9 years in 1982 to 79 years in 1983. Table 5.17 shows the age profiles of new tenants in 1992.

Anchor houses those people who are, by and large, on low incomes and highly dependent on state benefits. Table 5.18 shows that the largest number of tenants are on low income levels.

Table 5.17 Ages of new Anchor tenants January 1990 to June 1992

Less than 64	11%
65–69	17%
70–74	21%
75–79	24%
80–84	18%
85–89	8%
90 and over	2%

Source: NFHA CORE data

Note: This represents 7757 new tenancies over the period January 1990 to June 1992.

Table 5.18 Weekly household income of Anchor tenants January 1990 to June 1992

Less than £60	46%
£60–£79.99	20%
£80–£99.99	15%
£100–£119.99	9%
£120 and over	10%

Source: NFHA CORE data

Note: This represents income data on 6093 new tenancies over the period January 1990 to June 1992.

The accepted definition of poverty is 140 per cent of the income support levels. For a single elderly person the amount needed to live on, using this definition of poverty, would be approximately £80 per week. It is therefore a reasonable assumption, looking at the household composition of Anchor's tenants, that about 65 per cent can be defined as having poverty level incomes. The majority of these will be single elderly women.

Such images can be confirmed when the savings levels of new tenants are examined. The majority of Anchor tenants have no or limited savings – 72 per cent of tenants have savings of less than £6000. See Table 5.19.

Table 5.19 Savings levels of new Anchor tenants

None	29%
£1000–£1499	22%
£1500–£2999	12%
£3000–£5999	9%
£6000–£8000	3%
£8000–£15,999	14%
£16,000 and over	11%

Source: NFHA CORE data

Note: This represents savings data on 6217 new tenancies over the period January 1990 to June 1992

Anchor also provides support to older people through 34 Staying Put projects in England. These projects are part of a growing number of home improvement agencies across the country which are coordinated nationally by Care & Repair Ltd. Advice and support are given to older people in order to repair, improve and adapt their properties to enable them to remain in their own homes. On offer are: financial advice; technical and legal advice; assistance with raising finance through local authority grants and mortgages from building societies; and help with providing reliable builders and supervising the work.

Anchor has given help to over 25,000 elderly people over the past decade and has helped over 7000 people to have building work completed. The value of this is now in excess of £27 million. Anchor's projects have the following characteristics:

- around 50 per cent of clients are over 75;
- around 54 per cent of clients are single older people (of which

45 per cent are single women, 9 per cent single men) and the balance of 46 per cent are couples;

- around 70 per cent reported having health problems which restricted their mobility in their homes;
- 50 per cent were on low incomes (defined as less than £75 a week for single people and less than £125 a week for couples);
- 75 per cent had low savings (defined as less than £1000);
- around 40 per cent of clients lived in pre-1919 properties;
- around 7 per cent of properties lacked basic amenities.

POLICY IMPLICATIONS: HOUSING AND OLDER WOMEN

Much of the discussion presented here has focused on the contextual circumstances of older people, with the emphasis on trying to break down the generalisations about older people by government.

Older people are as diverse a group as any other section of society, if not more so. Yet, in spite of this, they have been labelled as one homogeneous group whose needs and aspirations are somehow similar.

This stereotyping of older people has damaging effects – it creates an image of older people with housing needs which are less than other groups' needs. Clearly, the 1980s have seen growing competition for scarcer resources among different groups in society with demonstrable housing needs.

This discussion does not seek to advocate more resources, as such, for older people. But there is a growing housing crisis which has seen increasing homelessness on a scale never experienced before; increasing mortgage arrears and repossessions across the country; and a massive shortage of affordable housing.

The presentation of the circumstances of older people, with a clear focus on older women as the major client group within this population, has attempted to demonstrate the complexity of old age and the dangers of gross generalisation which ultimately affect the perceptions of funders and providers of services to older people.

Older women find themselves, in many cases, to be one of the most disadvantaged groups in society. They tend to be on low incomes, often in poor housing, and opportunities for improving their quality of life are constrained by these circumstances, their age and the lack of affordable housing alternatives.

For older women who are now retired or who will retire in the next decade, income levels will remain low. This is reinforced by a number of trends. The marked income differentials between men and women during their working lives are perpetuated in the low incomes of women in retirement. The breaks from employment and the lower expectations of women in employment have created a situation where single or widowed women have low incomes. Often widowed women lose the income from a husband's pension. While there is some evidence that women's earning power is beginning to change within the current workforce, inequalities still remain. It is likely that it will be some time before women's incomes in retirement reflect the changes in earning power, and this will not happen until the next century.

Older women remain disadvantaged, on low incomes, highly dependent on state benefits, with little or no savings. Their incomes reflect their employment patterns and in particular their broad exclusion from the benefits of occupational pensions.

Much of the discussion in this chapter has focused on the housing circumstances of older people. The push towards owner occupation in the 1980s, as being the only housing answer, is clearly backfiring in the early 1990s. Homelessness, record repossessions and mortgage arrears, and a chronic shortage of housing have challenged the real benefits of a 'property-owning democracy' promoted by the present Government.

Currently, around 50 per cent of retired households are owner occupiers. This is likely to increase to between 65–70 per cent by the turn of the century. Many older people have lived in their property for over 20 years. Many will remain there until they die and, indeed, that would be what most older people want.

However, failing health will often mean that the property becomes more difficult to manage as people become older and frailer. The three-bedroom semi-detached house may become less manageable, more costly to repair and maintain, and a growing worry for older people.

Evidence from the experience of a number of providers is that the death of a husband can act as a strong 'push' factor in precipitating a move to different accommodation. Many older women, who move to either rented sheltered housing or bought retirement housing, do so in the aftermath of the death of a husband. For some the move to sheltered housing is a viable option; for others it is not.

Many of the government-inspired pointers about housing options in older age assume that the market will provide a range of options for older people to choose what is appropriate for them. Unfortunately, the reality is different.

Tenure has a powerful influence on determining the choice of housing for older women. Those in rented accommodation, either local authority, housing association or private sector, are broadly limited to moves within the rented sector. The real choice of moves is likely to be limited. There may be the option to move to sheltered housing, but this assumes that there is an adequate supply in the area and that the accommodation can be afforded. In many cases, the move to sheltered housing with rising service charges may not prove to be a viable option.

For those who are owner occupiers, there is potentially more choice but the reality may be a different matter. There has been considerable discussion recently about the equity value of properties, and how this will allow older people to 'trade down' into smaller accommodation and have capital left as well. This argument had its critics when house prices were at a high point, and now in times of a housing slump it seems even more suspect. Once again, a generalisation is being used as the answer to older people's housing, but there are a number of key flaws in the argument.

For a proportion of older women owning properties in the more affluent areas of the country, there may well be the opportunity to 'trade down' to smaller accommodation. However, this will depend on an affordable level of smaller accommodation being available. The choice may be either to move to smaller general housing or purchase retirement housing.

For many older women the 'trade-down' option may not be a viable choice. Trading down depends on the size and type of accommodation, the condition of the property and geographical location, all of which will ultimately affect the value. A two-bedroom terraced house in Oldham in poor disrepair may provide too low a value to move within the owner-occupied sector.

For those who are unable to move within the owner-occupied sector, there is growing evidence of larger numbers moving back to rented accommodation. This will mainly be to sheltered housing, either in the local authority or housing association sectors. By

1992 new lettings to Anchor sheltered housing from owner occupiers had reached 30 per cent. These were mainly single older
women. Once again, the available supply of affordable sheltered
housing in the location people want to move to may act as a
restraint on moving. Furthermore, there is growing evidence that
the affordability of sheltered housing is acting as a disincentive for
growing numbers of older women.

The other option for older women is to remain in their property
and use the services of a Staying Put or Care & Repair service. For
many older people, this has provided invaluable support and has
enabled older women to improve, repair and adapt their property
to allow them to remain living independently in their home. For
many the quality of their lives has been greatly improved.

Inequality remains a major concern for older women. For many
the disadvantages of employment opportunity have brought low
incomes in retirement. While the real value of the state pension
remains so undervalued from real need, many older women will
continue to live in considerable poverty. These low incomes and
savings levels ultimately affect real housing choices for older
women. Providers of housing and care services must be made
more aware of the needs of older women if this vulnerable group is
to see their housing needs addressed. For many older women, low
incomes, poor housing, increasing frailty, loneliness and social
isolation remain a reality. Government needs to act to tackle these
circumstances to bring real improvements to the quality of life of
older women.

REFERENCES

Bull, J. and Poole, L. (1989) *Not Rich: Not Poor – A Study of Housing
Options for Elderly People on Middle Incomes*, London: SHAC/
Anchor Housing Trust.
Central Statistical Office (1991) *Family Spending: A Report on the 1990
Family Expenditure Survey*, London: HMSO.
Department of the Environment (1988) *English House Condition Survey
1986*, London: HMSO.
Johnson, P . and Falkingham, J. (1992) *Ageing and Economic Welfare*,
London: Sage Publications.
Mackintosh, S. and Leather, P. (1992) *Staying Put Revisited*, Oxford:
Anchor Housing Trust.

FURTHER READING

Allen, I., Hogg, D. and Peace, S. (1992) *Elderly People: Choice, Participation and Satisfaction*, London: Policy Studies Institute.

Aughton, H. and Malpass, P. (1990) *Housing Finance: A Basic Guide*, London: Shelter.

Bone, M. (1992) *Retirement and Retirement Plans*, London: HMSO.

Bookbinder, D. (1991) *Housing Options for Older People*, London: London Concern England.

Bosanquet, N., Laing, W. and Propper, C. (1989) *Elderly People in Britain: Europe's Poor Relations?*, London: Laing & Buisson.

Butler, A., Oldman, C. and Greve, J. (1983) *Sheltered Housing for the Elderly*, London: Allen & Unwin.

Clapham, D. and Munro, M. (1988) *A Comparison of Sheltered Housing and Amenity Housing for Older People*, Edinburgh: Scottish Office.

Department of the Environment (1981) *Growing Older*, London: HMSO.

Fennell, G. (1986) *Anchor's Older People: What Do They Think?*, Oxford: Anchor Housing Association.

Fennell, G., Phillipson, C. and Evers, H. (1988) *The Sociology of Old Age*, Milton Keynes: Open University Press.

Fleiss, A. (1985) *Home Ownership Alternatives for the Elderly*, London: HMSO.

Fletcher, P. (1991) *The Future of Sheltered Housing – Who Cares?*, London: NFHA/Anchor Housing Trust.

Henwood, M. (1990) *Community Care and Elderly People*, London: Family Policy Studies Centre.

Jeffreys, M. (ed.) (1990) *Growing Old in the Twentieth Century*, London: Routledge.

Kiernan, K. and Wicks, M. (1990) *Family Change and Future Policy*, York: Joseph Rowntree Memorial Trust.

Laing & Buisson (1989) *Elderly Consumers in Britain: Europe's Poor Relatives*, London: Laing & Buisson.

Leather, P. and Wheeler, R. (1988) *Making Use of Home Equity in Old Age*, London: Building Societies Association.

McGlone, F. (1992) *Disability and Dependency in Old Age: A Demographic and Social Audit*, London: Family Policy Studies Centre.

Mackintosh, S., Means, R. and Leather, P. (1991) *Housing in Later Life: The Housing Finance Implications of an Ageing Society*, Bristol: School of Advanced Urban Studies, University of Bristol.

Middleton, C. (1987) *So Much for So Few: A View of Sheltered Housing*, Liverpool: Institute of Human Ageing, Liverpool University.

Oldman, C. (1988) *Leasehold Schemes for the Elderly: Ten Years On*, London: National Federation of Housing Associations.

Oldman, C. (1991) *Moving in Old Age: New Directions in Housing Policies*, London: HMSO.

Rolfe, S., Leather, P. and Mackintosh, S. (1993) *Available Options*, Oxford: Anchor Housing Trust.

Rolfe, S., Mackintosh, S. and Leather, P. (1993) *Age File '93*, Oxford: Anchor Housing Trust.

Taylor, H. (1986) *Growing Old Together: Elderly Owner-Occupiers and Their Housing*, Centre for Policy on Ageing: Policy Studies in Ageing No. 6.

Tinker, A. (1989) *An Evaluation of Very Sheltered Housing*, London: HMSO.

Wakefield Metropolitan District Council (1985) *Housing and Elderly People: Report of the Housing Needs of the Elderly Working Party*.

Whatmore, K. and Mira-Smith, C. (1991) *Eldercare in the 1990s*, London: National Carers Survey.

Wheeler, R. (1985) *Don't Move: We've Got You Covered – A Study of the Anchor Housing Trust Staying Put Scheme*, London: Institute of Housing.

Wilson, M. (1984) *Homes for Elderly People*, London: London College of Health.

Chapter 6

Black women and housing

Perminder Dhillon-Kashyap

Black women[1] in Britain as a group are as diverse as any other group of people – women from different countries, of different ages, sexuality, class, cultural, religious and linguistic backgrounds. However, the one common factor they share as consumers of housing – as Black women consumers of housing – is their experience of structural and subjective racism and sexism which determine their access to, as well as their choices in, the basic right of an adequate roof over their heads. This position is borne out in research, specifically carried out on race and housing and on the housing needs of Black women, as well as reflected in many recent reports and individual cases.

Black women's housing needs as a specific issue was first focused on in a coordinated manner by the Black Women and Housing Group formed in 1983 by Black women from the Greater London Council's Anti-racist Year Housing Group. At a local level, Black women organising autonomously were addressing the issue of domestic violence through setting up separate refuge provision as well as hostels for young women who were leaving their parental homes for a variety of reasons. Local Black women's groups were also opening up a dialogue with local authorities to get them to focus attention on Black women's housing needs which fell between the areas of race and gender policy. Similarly, Rao (1990) found that, although assumptions were made that Black women benefited from both policy areas, in fact each policy area, whether race or gender, gave scant consideration to the other, presumably assuming that the other policy forum would consider the issues jointly.

Therefore, to consider realistically the issues that affect Black women's access to and choice of housing provision – whether in the private or the public sector – one has to address the issues from both the race and the gender perspectives as well as take into consideration the economic position of Black women in Britain.

This chapter will outline the issues that Black women face in gaining access to adequate public sector and private sector housing provision. It will look at individual experiences of Black women as consumers of public sector housing in light of the policies and practices of local authorities in addressing the housing needs of Black women. It will conclude by looking at how Black women have organised autonomous provision for themselves in the face of State and community oppression and pressures.

BLACK PRESENCE IN BRITAIN

Over 3 million people from the Black and ethnic minority groups live in Britain today. That is 5.5 per cent of the total British population. Almost half are women. The history of migration by Black people to this country is well documented as governed by colonialism and imperialism (see Solomos, 1992). The Second World War saw the arrival of Black soldiers from the colonies to fight in the British Army or to help in the war effort. Post war, British subjects from the Colonies and the Commonwealth independent countries were encouraged to come to address the labour shortage. When the *Empire Windrush* arrived at Tilbury on 22 June 1948, with the first full shipload, people were accommodated in a Clapham air-raid shelter. But as Solomos (1992: 10) points out: 'It was during the period from 1945–62 that the terms of political debate about coloured immigration were established, leading to a close association between race and immigration in both policy debates and in popular political and media discourses.' Two themes predominated on policy. The first on immigration control and the second on the problems caused by too many 'coloured' immigrants in relation to housing, employment and crime. This is summed up in the phrase used by the Conservative candidate Peter Griffiths to electioneer in Smethwick in 1964 'if you want a nigger for a neighbour vote Labour'.

Apart from Uganda, from where Asian families migrated as family units, Asian[2] women on the whole migrated later than men, their arrival in Britain being governed by various factors like the timing of the main period of migration of men from India,

Pakistan and Bangladesh as well as the measures introduced by governments to reduce Black immigration. Asian women then came to join their menfolk who had migrated earlier. Migration of Indian women preceded that of women from Pakistan and Bangladesh. So although Asian women share many common experiences with other Black women, the settlement patterns produce different experiences for different communities. Therefore, the housing problems of long-established Sikh families as owner occupiers are different from those of Bangladeshi families in East London, who have been housed in poor quality council housing on estates where racial violence is a daily occurrence. Bangladeshi women, in particular, live totally isolated and in fear of attacks on these estates.

In comparison, as well as coming to Britain to join their menfolk, a substantial proportion of Caribbean and African women migrated independently in search of better prospects. Women were specifically recruited from Africa, the Caribbean and other former colonies to take up low paid jobs in industries like the National Health Service (Mama, 1989). In 1948, local selection committees constituting a centralised recruiting system had been set up in 16 countries including Nigeria, Sierra Leone, British Guyana, Trinidad, Mauritius and Jamaica. Consequently, more Caribbean and African women are breadwinners and heads of households.

In search of accommodation, early Black migrants to Britain were faced with notices proclaiming 'No coloureds here'. Consequently, many men were forced to rent 'beds' from other friends and relatives. Availability of scant accommodation gave rise to tremendous over-crowding. There are many stories of night shifters sharing the same bed as day shifters in crowded rented rooms. Forced to rent run-down properties or lodgings with repressive landlords, many Black settlers sought low-cost owner-occupied housing as the long-term alternative to initial rented lodgings.

From the late 1950s to the mid-1970s many Black people bought cheap houses in slum clearance areas or in neighbourhoods blighted by redevelopment. The mainstream building societies and banks would not lend mortgages on properties in these areas, which they regarded as bad investment. Mortgage exclusion zones were then set up which coincided with the areas where Black people could afford to buy. Because of this, property transactions were often managed and financed solely within the Black communities.

In some instances, communities could not afford to buy or rent and turned to community squatting as in the case of the Bangladeshi community in East London.

Thus, 'inner cities' and 'poor housing' came to be associated with racist sentiments against Black people which accused them of creating slums in the inner cities.

The same pattern exists today. The 1991 Census figures show that Black people have a greater tendency to live in the largest cities with 70.5 per cent living in the South East and West Midlands. Greater London alone accounts for 44.8 per cent. The figures also show that Black and ethnic minorities are least well represented in the higher status and the more rural parts of Britain.

As far as access to public housing is concerned, research studies during the 1970s and 1980s on race and council housing, conducted in a number of local authority housing departments (including Nottingham, Liverpool, Hackney, Birmingham, Bedford and Tower Hamlets), showed that Black applicants for council housing waited longer than White applicants and, once rehoused, received accommodation which was inferior to that given to White people.

Discrimination against Black people is still rife in council housing today. During 1992/3, the Commission for Racial Equality (CRE) conducted a formal investigation into Oldham Housing Department and monitored the non-discrimination notice served on Tower Hamlets Borough Council back in 1987 and Southwark Council in 1989. It also considered court action against Liverpool City Council for failing to comply with some requirements of a non-discrimination notice, but suspended this pending an agreement on a programme of work to be carried out by the Council. The CRE also published its report – *Racial Discrimination in Hostel Accommodation* – of the formal investigation into the allocation policies and procedures of Refuges Housing Association Ltd.

In 1990, the CRE formally investigated estate agency activity in Oldham, the third investigation of its kind, giving substance to the belief that racial discrimination is at least as widespread in owner-occupied housing as in rented housing.

Thus, a history of exclusion to decent owner-occupied housing, coupled with poor access to council provision, follows Black people, and particularly Black women, into the current situation.

LOCATING THE FACTORS WHICH GOVERN BLACK WOMEN'S ACCESS AND ALLOCATION TO HOUSING

Ginsburg (1992: 109) defines three kinds of process which often combine to produce the manifest racial inequalities in housing in contemporary Britain.

First:

> There are policy and administrative processes that have an indirect but fundamental impact on Black people's housing situation, notably government immigration and housing policies, the structure and workings of the labour market, and the interactions between 'race' and other major lines of stratification, particularly class and gender.
>
> (Ginsburg 1992: 109)

These processes Ginsburg calls structural racism because they are institutionalised in the socio-economic structure beyond immediate housing institutions.

Second: 'there are the policy and the administrative processes in local housing agencies, particularly local authority housing departments, building societies and estate agents, which have resulted in adverse treatment of Black people as compared to white people' (Ginsburg 1992: p. 109). This he calls institutional racism.

Third, there is overt racial prejudice and discrimination by key individuals which he describes as subjective racism.

STRUCTURAL RACISM

Racial harassment/racist attacks

The election which saw the victory of the first London local councillor in the Isle of Dogs, East London from the right-wing, fascist British National Party (BNP) in September 1993 was wholly fought on the issue of housing. Although the Asian population on the Island is only 12 per cent, the BNP put out propaganda which stated that the 'Asians were taking over our council houses'. Only 24 out of 135 council lets over the preceding 18 months had gone to

Asians. At the same time, there had been a dramatic increase in reported racist attacks. Given that one in ten racist attacks is reported, police figures for the Island and its vicinity put these at 53 which includes 16 violent attacks, excluding the attack on Quaddas Ali, the 17 year old still fighting for his life on a hospital ventilator.

Nationally, it is estimated that there is a racist attack every 28 minutes and that 10 per cent of all Black and ethnic minority households have been attacked in their homes. It is proven that women and children are the most vulnerable. A Home Office study in 1981 found that African/Caribbean people were 36 times more likely, and Asian people 50 times more likely, than White people to be attacked.

Thus racist attacks and propaganda – whether physical assault, verbal abuse or damage to property – are the primary factors which govern choice for Black families seeking council housing. From a survey of local authority housing departments titled *Living in Terror*, the Commission for Racial Equality (CRE) concluded that:

> [T]here are many, effectively 'no-go' areas which have acquired a 'name' for racial harassment and where members of ethnic minorities are afraid to accept offers of homes, should they be even offered them. We have examples of this from throughout the country. Far too often we hear of an 'outer-city ring' or whole sections of a local authority area where housing officers say they are reluctant to make offers to black people because of potential harassment. These areas often also contain good quality housing.
>
> (Commission for Racial Equality, 1987: 20)

An environment of overt racism renders Black women more open to racist attacks. Women are more likely to be home alone during the day, to take the children to school or to go out shopping. Single Black women living on their own or with children are even more vulnerable. A recent survey for South-Western Area Health Authority by Bristol University found that Black women, particularly South Asian women, suffered a high degree of loneliness and isolation because they were afraid of going out. Women stayed at home because they feared racist attacks and threats in public places.

Racist perceptions of Black women, with Asian women seen as

passive and vulnerable and African/Caribbean women as aggressive and sexually accessible, further exposes them to racially motivated sexual harassment. Women against Rape's report, 'Ask Any Woman' found that one in eight Black women suffered racist sexual assault. The psychological, emotional and physical effect that this has on Black women cannot be quantified.

Similarly, Rao (1990) in her survey of two London authorities found that most of the women interviewed complained of some form of racial harassment. Thirty-two per cent reported that they had been racially abused and an almost equal proportion had suffered physical attacks. A large proportion (23 per cent) had also experienced damage to property, especially broken windows, graffiti on the walls and cars being stolen or vandalised. Women also received racist telephone calls and letters. Nearly 45 per cent of the women had children or knew of children who had been harassed. The majority of the women did not report these incidents believing that nothing would be done. Only 40 per cent of those experiencing racial harassment reported to the authorities.

Both Rao (1990) and Mama (1989) found that Black women were extremely concerned about their personal safety, both on the streets and in their homes. The responses from the police and the local authorities to racist attacks/harassment were found to be inadequate.

Racial harassment is not exclusive to council housing in the inner cities. A recent study by the University of Warwick's ESRC Centre for Research in Ethnic Relations (Brar *et al.*, 1993) reveals patterns of racism in new towns such as Harlow which outstrip racism in the inner cities. Interviews with Black and ethnic minority groups showed that children suffered racist taunts and attacks in schools, women were spat at in shopping centres, youths were beaten up on the streets by racist skinheads and pubs were 'no-go' areas for Black men. At home, bricks through the windows and slashed car tyres were a common occurrence. Because of racist attacks/racial harassment, instead of the home being a sanctuary, for many Black women it is, or can be, a prison.

Immigration legislation

Structural racism also manifests itself in the way that immigration legislation works. Currently, three Black women are facing deportation having fled from domestic violence, and therefore contravening the 12-month rule which insists that a newly married couple must live together for 12 months. A fourth Black woman – Joy Gardner – accused of the same contravention, died while being arrested prior to deportation. Thus, for many Black women, escaping domestic violence not only means no offer of council housing but deportation as well. Under immigration regulations, if a woman has been sponsored by her husband to join him – which is the case for almost 99 per cent of all Black women – it is on the basis that he can accommodate her 'without recourse to public funds'. Leaving her sponsor involves contravening immigration legislation.

Not only can a local authority refuse housing to a Black woman escaping domestic violence but it should report its decision to the immigration authorities, following the decision by the Court of Appeal in March 1993. The Court of Appeal upheld the London Borough of Tower Hamlets' claim that it is entitled to investigate an applicant's immigration status, and that if it decides that an applicant is an 'illegal entrant' it may refuse to rehouse that person and her or his family under the housing homeless provisions. Further, the local authority should report its decision to the Home Office. So, in addition to refusing to house Black women if they leave their sponsors for some reason, local authorities will also 'police' their immigration status by informing the immigration authorities. The result could well be deportation.

This judgment has now given housing departments *carte blanche* to investigate immigration status, refuse housing on the basis of their conclusions and then inform the Home Office. The Commission for Racial Equality has already received numerous complaints from people who have been singled out for passport checks on the basis of skin colour, ethnic origin, nationality or accent and who have been required to *prove* an entitlement to services. Tower Hamlets has refused to rehouse families, in particular those from Bangladesh, who it has deemed to be 'intentionally homeless' for having left ancestral homes back in their countries of origin!

Many Black women are already apprehensive about approaching housing authorities for fear of the reaction they may receive. This Court ruling is effectively going to stop them. For Black women, leaving their sponsors and seeking accommodation in their own right is not a straightforward choice. It depends on their status under the immigration legislation which is not covered by the 1976 Race Relations Act.

National housing policies

Government policies on housing which encourage home owner-ship, coupled with the withdrawal of support – by various means – from council housing and council tenants, have further increased racial inequality in housing. These include cuts in subsidies to deal with the homeless, total stoppage in the building programme and the promotion of the right-to-buy schemes for council properties. The obligations which are to be placed on housing authorities to sell off large estates to Housing Action Trusts (HAT) and private landlords will further result in fewer properties being available to house the homeless.

According to the Chancellor's 1988 Autumn Statement, be-tween 1979 and 1988 public expenditure on housing was cut by 79 per cent in real terms. Council home sales, for example, since the 1980 Housing Act have been over 1.25 million. These have con-sisted largely of attractive suburban estates mainly populated by White households because of the racialised allocation policies of the past decades. The remaining housing stock, therefore, mainly consists of poorer quality, high-rise flats in run-down areas.

The 1988 legislation has further made housing association properties financially less attractive. At the same time, poverty is on the increase in the Black and ethnic minority communities. A report by the Child Poverty Action Group and the Runnymede Trust, titled *Poverty in Black and White – Deprivation and the Ethnic Minorities*, shows that unemployment among Black and minority groups is twice that suffered by White people. Ethnic minorities also tend to be in low paid jobs. If the recession continues, some Black and ethnic minority groups will be left out of the employment market altogether.

The report also shows that in housing the overcrowding figure was worse for Black and ethnic minority groups than for the White community. As Ginsburg puts it: 'Thus local institutional racism of the past combined with contemporary central government policy produce a powerful inegalitarian effect' (Ginsburg, 1992: 121). Similarly, a report by the London Housing Unit (Muir and Ross, 1993) shows that changes to housing policies have made it increas-ingly difficult to gain access to public housing, pushing the trend towards the private sector and to owner occupation. This puts women at a disadvantage as they are caught in the 'affordability' gap. For example, in Greater London, it was found that on average women full-time workers earn less than men and part-time women workers earn proportionately less than full-time workers of either sex. A vast majority (95 per cent) of part-time workers are women and women make up 24 per cent of unemployed people claiming benefits but 47 per cent of those seeking work. Seven out of ten families headed by a lone mother have a weekly income of £100 or less. There are no specific statistics available for Black women but the trends are similar. Consequently, many women, and therefore many Black women, will find themselves in urgent housing need.

Homelessness

Research by the London Housing Unit found that London's Black and minority ethnic households were up to four times more likely to become homeless than other households. Black women were twice as likely as White women to suffer long periods of homelessness. A survey of London's hostels in 1991 found that, on the night the survey was carried out, 40 per cent of hostel residents were women and nearly two-thirds of the residents under 24 years were female and black (Spaull and Rowe, 1992).

In Nirmala Rao's research (Rao, 1990), two-thirds of the Black women interviewed were homeless. The majority of the women were between 21 and 30 years old with White young women (16–20 years) particularly vulnerable to being homeless. Rao explains this as being due to the fact that many Black women are more likely, in the first instance, to seek support from friends and relatives during periods of homelessness, and less likely to approach housing departments for accommodation owing to the prevalence of racism. Young Black women are also more likely to leave home later than White women, given the norm of the extended family in Black communities.

Rao (1990) also found that only half the women who applied as homeless were, in fact, accepted as homeless in the two local authorities surveyed. Sixty per cent of homeless Black women were living with relatives and friends as compared to 42 per cent of White women. About 16 per cent of Black women and 18 per cent of White women were living in temporary bed and breakfast hostels provided by local authorities. Eight per cent of Black women were living in private rented accommodation.

Causes of homelessness

Women are more likely to experience homelessness than men because their access to housing is limited. Married women, in particular, are often dependent on their partners for income and therefore for housing, especially if they are mothers. In addition, as bearers and carers of children, women are less able to take full-time employment. Marital breakdown is therefore the primary cause of homelessness for women.

Rao (1990) found that the causes of homelessness among Black women included marital breakdown, domestic violence and loss of privately rented accommodation either because of unsatisfactory conditions or inability to afford high rents. Black women were also more affected by lack of support from friends and relatives, and as a result of overcrowding.

A third of the women had lost their homes following a marital dispute. A small proportion had lost their homes because of legal action taken against them for rent arrears and mortgage arrears, including rent arrears or illegal occupation of council housing. More White women than Black women were made homeless in the private rented sector. Racism and exorbitant rents usually precludes the private sector as a viable housing option for Black women.

Among the Black single homeless women, the causes cited in Rao's survey were disagreements with parents, conflicts with siblings, overcrowding, disputes over discipline, culture and other factors associated with a young person's life-style.

Many of the women suffered severe trauma as a result of their homelessness. Many suffered depression, isolation and loneliness. Poor living conditions also produced poor physical health. Women migrants from the Caribbean are twice as likely to have children who die during or shortly after childbirth (see Child Poverty Action Group and the Runnymede Trust, 1992).

Domestic violence

Referrals to London Women's Aid alone rose from 2019 to 6627 between 1984 and 1987; at the same time in London, there were only 300–400 bed spaces in women's refuges. An article in the *Guardian* (1993) revealed that a recent survey showed that one in ten women had experienced domestic violence in the past year.

Annually, Women's Aid Federation receives 100,000 calls on its help lines and 30,000 women and children use its refuges. Mama says: 'There is no doubt that significant numbers of these are Black women, even judging conservatively from the numbers of Black women in refuges at the present time' (Mama, 1989: 101).

With the breakdown of a relationship, the question of housing becomes crucial for the woman. The loss or reduction in income makes it difficult for a woman to rent in the private sector, leaving local authority housing as the only viable option, an option which many women are finding extremely hard to obtain.

Many women, especially Black women whose first language is other than English, are not aware of the provisions of public housing. Rao (1990) found that only four women in her sample knew about their rights to accommodation. She also found that many Asian women were reluctant to approach agencies for fear of being identified. Being stigmatised for leaving a marital home can be more harrowing for Black women who have little or no support in terms of their own family/relative networks in this country, having to rely on the husband's family and relatives.

The 1983 Parliamentary Committee on Domestic Violence made recommendations to local authorities on housing women suffering from domestic violence. Mama observed that, 'yet in 1989 very few of the recommendations have been met in real terms' (1989: 200). Rao also found that, while Southwark Council had a policy on recognising the housing needs of women leaving domestic violence, 'there appear to be considerable differences in interpretation between districts and a consistent borough-wide approach is lacking' (Rao, 1990: 35).

Structural racism, combined with patriarchal oppression are then the context in which statutory authorities perceive and re-spond to the housing needs of Black women.

However, being aware of, and having access to, council housing are two different things. A Black woman's immigration status can be one of the determining factors in her entitlement to council housing.

INSTITUTIONAL RACISM AND SUBJECTIVE RACISM

A survey by the Association of Metropolitan Authorities (AMA, 1985) found that, out of the 61 authorities which responded to the questionnaire, only 24 were keeping ethnic records for transfer applications, waiting list applications, nominations to housing associations, housing renovation grants, housing benefit and homeless persons. These statistics were not broken down for Black women. In addition, only 22 authorities kept separate records on racial harassment, 14 being London boroughs/Greater London Council. Although records were kept, monitoring was not regular or effective.

It was also revealed that, although some authorities were unable to provide adequate figures, other authorities did recruit a small number of specialist staff catering for the needs of Black and ethnic minority communities. Less than 29 authorities ran any courses for staff on race and/or the housing needs of Black and ethnic minority communities. Literature was translated into other languages by 21 authorities. Seventeen authorities held regular meetings with representatives of Black and ethnic minority communities.

In Rao's (1990) survey, only Wandsworth Council had monitoring systems but this again did not include analysis by gender. Statistics showed that Black homelessness was on the increase; offers and lettings to Black homeless families were far below the total approaching the Housing Department; and Black people on average spent longer in temporary accommodation than White people. Southwark Council had not introduced monitoring, but in 1985 it adopted an overall policy called 'Equality for Women' which examined women's, and particularly Black women's, experiences of discrimination and made recommendations. Implementation was lacking. The Housing Department had also drawn up a positive action plan in 1988 but the extent of implementation was not known.

Research by the Commission for Racial Equality in 1990 showed that 28 local authorities surveyed had specific policies and/or strategies for race equality. A recent survey by the Local Government Management Board (March 1993) showed that 82 per cent of authorities had adopted an equal opportunities policy. But translating these policies into practice for Black women seeking council housing is another matter.

ACCESS TO COUNCIL HOUSING

Housing the homeless

If she is not already a council tenant, access to local authority housing provision for a homeless woman depends on the woman being accepted as 'priority' under the homeless category. The legal definition of homelessness is contained in the Housing (Homeless Persons) Act 1977 which is now part of the 1985 Housing Act. The Act defines categories of homeless people who are considered to be in priority housing need. These include those with dependent children. Single homeless people are only considered if they are regarded as being 'vulnerable' because of old age, mental illness and handicap, or physical disability or special need. Priority is also based on whether the woman is vulnerable, e.g. pregnant or with a mental illness or with custody of the children. Domestic violence is not an automatic ground for being accepted as homeless, although recently a number of local authorities have tried to devise policies around this issue. But Black women's experiences show that these policies have not benefited them at all.

Mama (1989) examined the experiences of Black women who had suffered domestic violence and relationship breakdown when they approached local authority housing departments. Given that many Black women face cultural/family pressures to the point of being stigmatised by family and relatives for leaving violent partners, when these women do approach council departments they are in dire need. Mama (1989) states that, with the recent adoption of anti-racist and domestic violence, policies one may expect a more sympathetic response from local authorities towards Black women who have experienced domestic violence. This is not so. For many Black women, leaving a violent home may well mean embarking on a long ordeal of racism, homelessness, near destitution as well as fighting immigration laws.

Many of the women interviewed reported insensitivity and hostility from officers, even though they did not articulate these experiences as racist. For example, an Iranian woman, having been told to go back to Iran, did not describe this as racist. However, many women later came to connect their experiences with the fact that they were not white. Mama (1989) describes how one pregnant woman, escaping a violent relationship, went to a housing department virtually every week and then for one year after her child was born only to be offered a fourth floor flat in a 'really filthy' place. Many Asian women, approaching homeless units, have been told that they could not possibly have left their violent husbands as their culture does not allow it – a corollary of the 'virginity' tests that the State inflicted on Black women coming to this country back in the 1970s and 1980s. In order to prove their claims of being genuine wives-to-be, Black women had to undergo 'virginity' tests as part of an immigration check, because 'their cultures demanded that women should be virgins before marriage'!

MacEwen (1990) shows how local authorities have interpreted the homeless legislation – sometimes through the courts – in a way that discriminates against Black people.

Once accepted as homeless, living in local authority temporary accommodation is the most distressing experience for women, not least because of the quality of the accommodation. Provision for homeless women includes refuges, hostels, hotels, short-life property, and bed and breakfast. Many Black women describe these places as dark and dingy with no adequate facilities. Many women suffer sexual and racial harassment from other residents as well as managers, and many with children describe how they have to leave the temporary accommodation between certain times and walk the streets.

For single homeless Black women, local authority housing is often the only available form of tenure. However, local authorities have no legal duty to house single homeless people unless they are deemed to be 'vulnerable'. Interpretation of being 'vulnerable' can be very subjective and does depend on the investigating personnel. In Rao's research (Rao, 1990), single Black women felt that being Black affected their chances of obtaining accommodation. They often found officials to be rude and abusive and had to face a lot of racism. If single women were offered accommodation, it was bed and breakfast accommodation and that for a limited time.

Many homeless Black women wait longer to be rehoused. Again, in Mama's research (Mama, 1989) she found that both Black and White workers in mixed refuges observed that it was often the Black women in the refuge who waited the longest time to be rehoused. Lewisham refuge actually monitored the rehousing of Black and White women and found racial differences which they submitted to the local authority.

Experiences of Black women as compared with White women show that discrimination in allocation is still prevalent. Both Rao

(1990) and Mama (1989) found that, compared to White homeless women who had a higher rate of refusal of the first offer of accommodation, Black women were not given any choice as regards the suitability of the preferred accommodation. Consequently, many Black women were allocated flats which were subject to vandalism, and in areas where the women were exposed to violence and/or abuse. A large number of Black women were offered (48 per cent as compared to 26 per cent) poor quality accommodation, often in the least popular areas. Almost three-quarters of Black women, compared with half the White women in the sample, were allocated flats, while less than 1 per cent of Black women were offered terraced or semi-terraced properties. These findings are consistent with research by the Commission for Racial Equality and the Policy Studies Institute on the housing position of Black people.

Different experiences for different communities

ESRC data show that, compared to Asian women (except for the Bangladeshi community), women from the African/Caribbean communities are proportionally better represented in council accommodation. In the past, this has been interpreted as arising from class differences between the two communities as well as from the linguistic advantage of the African/Caribbean communities over the Asian communities. But, in addition to the different migration patterns of the two communities, Peach and Byron (1993) show that this is because of a complex interplay between class, gender and racism.

There is a greater proportion of female heads of household in the African/Caribbean communities. During 1986–9, 75 per cent of White households were male-headed while 25 per cent were female-headed. For Caribbean households, the figures were 59 per cent and 41 per cent. While a third of Caribbean male-headed families were in council housing, nearly two-thirds of female-headed families were in this tenure. Many of these were single mothers with children. This, Peach and Byron observe, relates to class background as well as the fact that female-headed families are the norm in the Caribbean. Similarly, the migration patterns of many women from Africa and the Caribbean, who came as workers in their own right, contribute to this. Again, statistics also show that more women from this group are allocated flats in high rise buildings when compared to White women from the same group.

Proportionally, more women from Indian and Pakistani backgrounds are in the owner-occupier and private rented sector. This is due more to families and communities pooling together to buy homes on migration to this country (as would be the norm in extended families) than a class difference between the indigenous and the Asian families. Migration patterns and class differences are also responsible for the Bangladeshi communities being located proportionately more in public sector housing. But, as research shows, because of unemployment, there is a growing demand from all Black communities for affordable public sector housing.

Owner occupation does not mean better quality of housing, given the areas in which Black families were forced to buy. Many Black women still find themselves in run-down properties because of how racism governs their choice. They also find themselves in crowded circumstances owing to the size of the properties. The experiences of Asian women show that, when they have applied for local authority housing on grounds of over-crowding, after initial disbelief ('your culture says you shouldn't leave extended families'), they are blamed for creating the over-crowding 'by sticking together'. Indeed, when Asian women joined a housing campaign group, Southall Action on Housing in West London in the early 1980s, they squatted in the four-bedroom properties that the local council had just built for sale. The Chair of Housing spoke on the local radio in defence of the sales and condemned Asian families for creating the over-crowding. Accusing the victims of creating the problem and then taking away any means of a solution seem to be the norm in council housing provision.

Local authority policies

Analysis of the Greater London Council's (GLC) housing allocation policies in Tower Hamlets (see Mullins, 1988 and Forman, 1989) showed that official judgements were being made about the suitability of Asians and non-Asians in matching vacancies to requests. The Bangladeshi community had virtually been excluded from the new GLC housing and had suffered for a long time from allocation to deck-access flats based on a policy of dispersal to estates beyond Spitalfields. Similarly, Henderson and Karn (1987) found, in their study of Birmingham housing policy and practice, that there were six policies which put Black families at a disadvantage:

1 The requirement for applicants to be resident in the city for at least five years, which was reduced to two years in 1977.
2 The disqualification of applications from owner occupiers.
3 The disqualification of applications from single people of most ages.
4 The disqualification of applications from unmarried couples, unless cohabitation had lasted five years for childless couples and two years for couples with children.
5 The rejection of applications from joint families.
6 The policy of dispersing Black families within the housing stock, operating from 1969 to 1975.

In addition to formal policies, Henderson and Karn (1987) also found that discretion by housing officers at all levels further disadvantaged Black people. As shown earlier, racist and sexist assumptions about Black women, coupled with no specific policies addressing their needs, further disadvantage them.

Thus, institutionalised racist processes, together with racist attitudes, class prejudices and gender stereotypes, is the context in which council housing allocation takes place for Black women.

Private/owner-occupied sector

As mentioned earlier, the discriminatory practices of the mainstream lending institutions had forced Black people to buy particular properties in specific areas. The Commission for Racial Equality's formal investigation of an estate agent in Oldham in 1989 (Commission for Racial Equality, 1990) showed that the firm recommended White areas to prospective White purchasers and Asian areas to prospective Asian purchasers. They also accepted instructions from White vendors to deter Asian buyers and only offered mortgage facilities to White clients. The CRE found that two other estate agents in Oldham were practising in similar ways. The CRE believes that such discrimination may not be uncommon, given that there have been three formal inquiries into estate agents.

The experiences of Black women in the owner-occupied sector, as in the public sector, are linked to subjective and institutional racism.

Calculations by the London Housing Unit in their report *Housing the Poorer Sex* (Muir and Ross, 1993) show that, given the economic status of women, only 19 per cent of women can afford to rent a bedsit in London. In addition, many housing associations are forced by legislation to charge market rents, therefore excluding this area as a viable option. Hence, many women – and many Black women – are left to look to council housing provision or share with friends and relatives, rendering the latter group invisible as far as any assessment of their housing needs is concerned.

Addressing the imbalance

Black women are adversely affected by national government policies which seek to shift the balance from the public to the private sector. Given that many women rely on their relationships for access to owner occupation, women on their own will not be able to afford housing in this sector because of their lower income. Many Black women work in the unorganised sector or at home and the proposed abolition of the Wages Council will further affect their chances of earning a decent wage, and therefore their ability to afford housing on their own. Nor will they be able to pay exorbitant rents in the private sector or in the housing association sector, from which they may easily be evicted under the 1988 Housing Act. Given these factors, the number of Black women will continue to grow in the group with urgent housing need, and they will be forced to seek council housing.

Public sector housing is, therefore, an important source of housing for many Black women. It is essential, then, that local authorities should adopt and adapt policies to assess and meet the housing needs of Black women. Some recommendations for local authorities, which would benefit all women, should include:

• developing an imaginative, comprehensive policy on housing women by researching and consulting with users and women's groups, and taking into consideration the specific needs of and provision for Black women. Policies on race and gender equality should be brought together to ensure that Black women's needs receive the priority and recognition which have been eclipsed so far by treating the two areas separately.

The policy should also address affordability, women-only

housing and support provision for women, e.g. childcare, health, cultural needs. Safety should be a high priority for existing and future council housing.

Black women should also be involved in all stages of decision-making from planning to building to regular monitoring and review. The guidelines on planning by the CRE should be followed.

Housing provision should be based on need. The new focus on tenant participation and customer charters are a good building block for this kind of effective consultation and involvement.

Housing needs for single women should be given a high profile and recognised in policies and provision.

• a comprehensive awareness/training programme for all housing personnel to eliminate stereotyping and racist and sexist attitudes/views. Discriminatory practices should be identified and removed. Indeed, local authorities are bound under the race and sex discrimination legislation to deliver all their services in a non-discriminatory way. The CRE's codes of practice for rented and non-rented housing provision are important guides in this respect.

Staff should be trained to deal with racial and sexual harassment, and more women including Black women should be employed to deal with customers. Input from local Black women and groups should be actively encouraged on an ongoing basis to provide the 'dipstick' for checking out policies.

• comprehensive monitoring of all housing services – including breakdowns for Black women – should take place with regular review and implementation to address discriminatory practices and gaps.

• comprehensive policies and action on racial and sexual harassment to be adopted by actively involving tenants, Black organisations, women and Black women. A 'victim'-centred approach should be adopted, and the local authority should work with other key statutory and voluntary bodies in a coordinated multi-agency way to ensure that all incidents are dealt with and monitored. Victims should receive appropriate support and perpetrators should be dealt with promptly.

Tenancy agreements should incorporate clauses on racial and sexual harassment.

• the provision of adequate homeless accommodation, regulated by the council's own practices in meeting the housing needs of Black women. Bed and breakfast accommodation should be used as a last resort and never for women with children. When used, it should be in line with the Department of Environment's Code of Guidance accompanying the 1977 Housing (Homeless) Persons Act.

- the removal of discriminatory policies and practices such as residential qualifications, dispersal of Black families (which still takes place covertly), disqualification of joint families, owner occupiers and co-habiting couples.

Local authorities should examine how they are interpreting the homeless legislation and how they seek to involve the judiciary in these interpretations. The Greve inquiry (Greve, 1985) stated that the operation of the Homeless Act was characterised by 'ethno-centricism and racially discriminatory practices'. There is a role here for the CRE and the Equal Opportunities Commission to lobby for greater power in these areas.

- lobbying central government to re-initiate a building programme based on housing need – moving away from the traditional focus of a nuclear family to more imaginative housing schemes. The selling of local authority housing stock must be halted.

- providing adequate information and publicity by producing translations of written material as well as interpreters in local languages. More Black women should be employed in key housing areas to promote equality of service provision.

In addition to local authorities adopting and adapting policies and practice, courts need to be guided as to how they are interpreting housing legislation. Their interpretations, as shown in the Tower Hamlets case, have a directly discriminatory effect on Black women.

Providing a viable housing option for women in the private sector depends on improvements in several areas. Discrimination against women in the labour market should be tackled. This could be done through the existing legislation on race and sex discrimination. But, as the two reviews of the 1976 Race Relations Act have shown, the powers of the Commission for Racial Equality are limited as far as enforcing change is concerned. The Race Relations Act and the Sex Discrimination Act should be extended.

Even then, only full-time women workers will benefit. Part-time female workers and women working in the unorganised sectors need more protection under the law, not less. The abolition of the Wages Council and the increasing anti-worker legislation should be halted. More protection for women workers should be brought in, in line with some European Community legislation.

For many women their employment pattern is governed by their caring or parental responsibilities. Until society recognises this role and provides financially for women in this position, women will always be at the lower end of the economic ladder, having their chances of affording housing in their own right further reduced.

In the current situation, central government should make a significant financial contribution in order to meet the housing demands of the majority of women. As the report *Housing the Poorer Sex* (Muir and Ross, 1993) recommends, this should take the form of a sufficient housing subsidy to housing providers channelled through local councils and housing associations. Further, homes should be allocated on the basis of housing need.

As Ginsburg puts it, 'Countering the overwhelming effects of structural racism, particularly the effects of government policies, is an equally formidable task' (Ginsburg, 1992: 130). For Black women, these range from immigration and welfare services to employment policies, as well as gaps in policies on racism.

Black women's struggle for autonomy

Whilst facing state and patriarchal pressures, Black women have organised their own particular struggles around issues. Black women have fought hard for the right to organise independently of the White women's movement, which made assumptions about Black women's oppression by Black men and did not necessarily address the issue of racism. Black women have also fought for their autonomy within Black liberation groups which were slow to take on gender perspectives because 'the struggle would be divided'. There are countless campaigns – too numerous to catalogue here – around employment, immigration, housing and domestic violence that Black women have initiated and been involved in (see Braham *et al.*, 1992). Suffice to say that, while local authorities have been developing policies around race and gender, Black women have been organising their own housing provision.

The first two Black women's refuges were set up by groups of Asian women in the late 1970s in the London boroughs of Brent and Southwark. The need for an autonomous Asian women's refuge, run by Asian women, arose out of the experiences that Asian women had of using the Women's Aid refuges. Many women encountered racist assumptions and views – from the other women using the refuge and from workers. Many experienced harassment and extreme isolation in these refuges. The experience of being subjected to violence at the hands of men can serve to unite women, but it can also lead to racist assumptions about Black men.

Asian women then campaigned hard to set up autonomous refuges amid great opposition from local communities, local women's refuges and local funding authorities. The funding authorities had to be convinced that Asian women's refuges were better placed to support and understand Asian women in terms of family ties, customs and traditions as well as language requirements. Tremendous battles had to be fought with 'community leaders' who saw the setting up of these refuges as encouraging family and community breakdowns. They also saw the women who were setting up these refuges as 'immoral, loose women leading other women astray'. Black women also worked with the local Women's Aid refuges to persuade them to recognise Black women's needs and to assure them that a united front would be presented to the funding authorities.

Given that specific funding for minority groups is based on 'special needs', largely around language and customs, women from the African and Caribbean communities had a harder struggle to get the funding authorities to recognise their particular needs. This is reflected in the numbers of refuges that exist today. There are 14 Black women's refuges in Britain, with two specifically catering for women of African and Caribbean origin. With the increase in Black women workers in Women's Aid refuges, a Black women's group was formed within the National Women's Aid Federation (NWAF). Black women's issues are now being addressed by NWAF through their policies and practices, and many Black women's refuges are affiliated to it.

In addition, because Black women have to fight on numerous fronts, many Black women's refuges have led historic struggles. These include campaigns on deportation, housing, welfare services, funding and domestic violence, both in the local communities and at national level. The Kiren Aluwahlia case is one such campaign where Black women managed to get the courts to set a precedent on 'provocation'.[3]

Housing provision around domestic violence is not the only issue that has concerned Black women. Many Black women are actively involved in setting up autonomous housing provision – both for themselves as women and for their communities. Statistics from the Federation of Black Housing Associations (FBHO) reveal that there are 120 Black housing associations with at least 29 specifically dealing with Black women's housing needs.[4] These range from the permanent provision of housing for single Black women, young women and elderly women to half-way homes.

CONCLUSION

In summary, there are countless issues around Black women and housing. First, the position of Black women in terms of the oppression they face in society as a result of their class, race and gender needs to be recognised in addressing their housing needs. The issue of access and affordability, both in the public and the private sector, needs to be addressed. Local housing authorities need to incorporate the specific housing needs of Black women in their policies and practices. These include providing adequate information and housing provision, and removing discriminatory practices on acceptance, allocation and quality of housing provision. It also means changing oppressive attitudes and addressing the issue of safety. Black women should also be involved in the planning and provision of council housing. Local authorities should lobby for change in central policies which are moving away from public to private housing provision.

Changes in the private sector are also dependent on changes in the labour and housing markets which have an impact on whether women can afford to rent or buy in the private sector. Central government policies need to address imaginatively the economic position of women, especially in view of their caring role in society. Affordability of housing in the private sector is increasingly becoming an issue in the current recession, and the need for subsidies should be considered by the Government.

As Cassidy and Gohil (1989: 19) in the special issue of the FBHO's magazine *Black Housing* on Black women and housing, say about local authority and housing association provision: 'if gender oppression is to be regarded as a serious issue . . . then the parameters of Equal Opportunities need to be reassessed and extended to encompass wider definitions.'

NOTES

1 'Black women' is used as a political term for women originating from, or having their roots in, Africa and Asia who experience racism in the British context.
2 The term 'Asian women' here refers to women originating from the Indian sub-continent.
3 The Kiren Aluwahlia campaign was run by Black Women, c/o Southall Black Sisters, Norwood Road, Southall, Middlesex.
4 See the *National Directory of Black Housing Organisations* (1993) published by the Federation of Black Housing Organisations, 374 Gray's Inn Road, London WC1.

REFERENCES

Association of Metropolitan Authorities (1985) *Housing and Race: Policy and Practice in Local Authorities*, London: AMA.
Braham, P., Rattsani, A. and Skellington, R. (eds) (1992) *Racism and Anti-Racism: Inequalities, Opportunities and Policies*, London: Sage in association with the Open University.
Brar, H., Martin, P. and Wrench, J. (1993) *Invisible Minorities: Racism in New Towns and New Contexts*, Warwick: University of Warwick Centre for Research in Ethnic Relations, Monograph 6.
Cassidy, M. and Gohil, A. (1989) 'Black women and housing', *Black Housing*, Vol. 5, No. 6: July, Federation of Black Housing Organisations.
Child Poverty Action Group and the Runnymede Trust (1992) *Poverty in Black and White – Deprivation and the Ethnic Minorities*, London: Child Poverty Action Group and Runnymede Trust.
Commission for Racial Equality (1987) *Living in Terror*, London: CRE Publications.
Commission for Racial Equality (1990) *Racial Discrimination in an Oldham Estate Agency*, London: CRE Publications.
Forman, C. (1989) *Spitalfields – A Battle for Land*, London: Hilary Shipman.
Ginsburg, N. (1992) 'Racism and housing concepts and reality', in P. Braham, A. Rattsani and R. Skellington (eds) *Racism and Anti-Racism: Inequalities, Opportunities and Policies*, London: Sage in association with the Open University.
Greve, J. (1985) *Homelessness in London*, Interim Report, Leeds: University of Leeds.
Henderson, J. and Karn, V. (1987) *Race, Class and State Housing*, Aldershot: Gower Press.

Local Government Management Board (1993) *Equal Opportunities in Local Government*, Birmingham: Local Government Management Board and Society of Chief Personnel Officers.

MacEwen, M. (1990) 'Homelessness, race and law', *New Community*, Vol. 16, No. 4: 505–21.

Mama, A. (1989) *The Hidden Struggle*, London: London Race and Housing Research Unit.

Muir, J. and Ross, M. (1993) *Housing the Poorer Sex*, London: London Housing Unit.

Mullins, D. (1988) 'Housing and urban policy', *New Community*, Vol. 17, No. 1: 114–23.

Peach, C. and Byron, M. (1993) 'Caribbean tenants in council housing: race, class and gender', *New Community*, April, Vol. 19, No. 3: 407–23.

Rao, N. (1990) *Black Women in Public Housing*, London: London Race and Housing Research Unit, Black Women in Housing Group.

Solomos, J. (1992) 'The politics of immigration since 1945', in P. Braham, A. Rattsani and R. Skellington (eds) *Racism and Anti-Racism: Inequalities, Opportunities and Policies*, London: Sage in association with the Open University.

Spaull and Rowe (1992) *Silt-up or Move-on?*, London: Single Homeless in London.

Chapter 7

Young women and homelessness

Anne Douglas and Rose Gilroy

> Shelter is one of the basic human rights. The expectation that the advent of the welfare state would eliminate homelessness has not been realised.
>
> (Women's National Commission, 1983)

The welfare state and the housing policies grafted within it are predicated upon the assumption that the nuclear family of mum, dad and children is the norm, with other groups such as young single people, childless couples, one parent families and old people as formative, transitional and residual stages of the family. The needs of these other groups can only be met if they do not prejudice the position of families (Brailey, 1985). This chapter examines the housing position of women who are homeless and for various reasons are not part of the nuclear family.

The core of this chapter is an examination of young women's homelessness in south-east Northumberland. This overturns the concept that homelessness is only a big city problem and highlights the work of a small voluntary organisation (Housing Aid for Youth) faced with helping young people under 26 – a group that central government has targeted in social security and employment policy.

THE HOMELESS LEGISLATION

The chapter begins with a study of the available statistics. Good statistics should underpin any social policy but the debate often seems to be about the figures themselves. John Greve and Elizabeth Currie summarise the debate as follows:

There is no universally accepted definition of homelessness. What constitutes 'homelessness' and how many people are homeless is a debate which has been running for thirty years or more.

The most important obstacle to the achievement of a consensus is that whatever definition is accepted by public authorities carries with it an implied responsibility to initiate or support action to reduce homelessness and, thus, to commit public resources to that end.

A range of definitions has been proffered. The narrower definitions focus on 'rooflessness' – the total lack of anywhere to live – while broader definitions include judgements about the quality of accommodation and about its social functions.

(Greve and Currie, 1990: 28)

Statistics are derived in the main from the operation of the Homeless Persons Act 1977 (now enacted as Part III of the Housing Act 1985). The operation of the Act varies considerably from authority to authority (Niner, 1989) depending on their policy and practice in interpreting the Code of Guidance, their political ideologies, the resources available to them and the priorities in the use of these resources (Greve and Currie, 1990).

Broadly the law states that a local authority must satisfy itself on three points:

- Is the applicant homeless or threatened with homelessness?
- If so, is the applicant in priority need?
- Is the applicant intentionally homeless?

In addition, in cases where homelessness has been proved as well as priority need and there is no intentionality, the authority must determine whether the applicant has a local connection with the authority to which she or he has applied and what to do if there is no local connection (DOE and DH, Welsh Office, 1991).

Problems arise at every stage of an inquiry. On determining 'homelessness' the Code of Guidance advises that an authority must consider reasonableness in respect of over-crowding, type of accommodation, threats of violence or actual violence, and security of tenure. Niner's research (1989) indicated that issues such as the condition of the property and security of tenure may only be grounds for adding that individual or household to the waiting list, not declaring them homeless. Carol Sexty (1990) finds that councils do not accept harassment on the grounds of sexuality as grounds for leaving home, so gay men and lesbians leaving violent situations are in danger of being classed as intentionally homeless and refused accommodation. The Code of Guidance still has nothing to say about sexuality or those who are victimised because of AIDS/HIV.

The biggest minefield lies in assessing priority need. The legislation specifies the following as priorities:

- a pregnant woman or a person with whom a pregnant woman resides or might reasonably be expected to reside;
- a person with whom dependent children reside or might reasonably be expected to reside;
- a person who is homeless or threatened with homelessness as a result of an emergency such as flood, fire or other disaster.
- a person who is vulnerable as a result of old age, mental illness or handicap or physical disability or other special reason, or with whom such a person resides or might reasonably be expected to reside.

Table 7.1 shows that parents are clear winners, while those in other groups are more likely to be losers or, anticipating little help, decide not to subject themselves to the test of law (Table 7.2).

Table 7.1 Homeless households found accommodation by local authorities by priority need category

Priority need category (%)	1986 000s	1989 000s	1991 000s
With dependent children	66	68	65
Pregnant	13	13	13
Vulnerable because of:			
Old age	7	6	4
Physical handicap	3	3	3
Mental illness	2	2	3
Other reasons	6	6	9
Homeless in emergency	3	2	1
All categories	108.9	130.0	160.1

Source: Central Statistical Office, 1993

Table 7.2   Local authority inquiries under the homelessness legislation:
by outcome

Households applying as homeless	1986 000s	1990 000s	1991 000s
Accepted			
– in priority need	109	151	160
– not in priority need	10	14	9
– intentionally homeless	3	5	6
Given advice and assistance only	59	86	81
Found not to be homeless	68	83	88
Total inquiries	249	339	344

Source: Central Statistical Office, 1993

The position of the very young (the under-18s) is particularly
interesting, since some will be among the 14,400 found to be
vulnerable for 'other reasons'. Many others may be among the
81,000 given advice and assistance only.

A new Code of Guidance accompanying Part III of the Housing
Act 1985 was issued in 1991. This specifically mentions young
people and the circumstances in which they may be classified as
'vulnerable'. It stipulates that:

> Young people (16 or over) should not automatically be treated
> as vulnerable on the basis of age alone. Young people could be
> 'at risk' in a variety of ways. Risks could arise from violence or
> sexual abuse at home, the likelihood of drug or alcohol abuse or
> prostitution. Some groups of young people will be less able to
> fend for themselves than others, particularly for example: those
> leaving local authority care; juvenile offenders . . . ; those who
> have been physically or sexually abused; those with learning
> difficulties and those who have been the subject of statements of
> special educational need.
>
> (DOE and DH, Welsh Office, 1993: 6.13)

These statements are not intended to be seen as a complete list but
no special consideration is given to the position of young women.
Case law is useful here as in the statement by Lord Ross in Kelly
vs. Monklands District Council (1985) that:

When you find a girl of 16 with no assets, no income and nowhere to go and who has apparently left home because of violence, I am of the opinion that no reasonable authority could fail to conclude that she was vulnerable.

Relatively few young people appear to be accepted as having any priority status under the homeless legislation. Jane Dibblin's research (1991) revealed that only 17 per cent of local authorities will usually accept under 18s as homeless. In 1990–91, 23 per cent of Housing Aid for Youth's (HAY's) clients were considered priority cases; 60 per cent of these were female clients and made up 28 per cent of all women clients of the organisation. The grounds for priority status were predominantly pregnancy and physical or sexual abuse in the home. The remainder of HAY's homeless clients only had a statutory right to advice and assistance, which at best meant obtaining a list of bed and breakfast places. Such lists tend to stipulate lower age limits and whether people in receipt of benefit will be accepted – all of which can be used as gate keepers by those operating bed and breakfast places.

It was hoped that the introduction of the Children Act 1989 in October 1991 would enable more young people to secure accommodation. Under this Act Social Services have a duty to provide accommodation to any child in need – this covers any person up to and including those aged 17 who fall within certain defined categories (Children Act, 1989, s 17). These young people must be provided with accommodation if one of four situations applies to them, including provision of accommodation where the welfare of the young person will be seriously prejudiced if accommodation is not provided (s 20 [1] and s 20 [3]). Helen Kay writing in 1992 suggested that the combination of the two pieces of legislation should enable more 16 and 17 year olds to secure permanent accommodation. The Children Act Housing Action Group (CAHAG) (1992) argued that:

> the definition of 'child in need' and 'seriously prejudiced' as it relates to homeless 16 and 17 year olds is broader than the definition of vulnerability under the Housing Act 1985 Part III. However, once a young person has been accepted under s20 [3] (Children Act), we would argue that they then come under the definition of vulnerability under the Housing Act Part III.

> (CAHAG, 1992: 2)

While there is no formal overlap between the two pieces of legislation, CAHAG's argument is given some support by the Code of
Guidance to the Children Act which stipulates that 'vulnerability'
under the Housing Act and 'serious prejudice' under the Children
Act might be expected to arise in similar circumstances and that
housing authorities will need to have a regard to a social services
authority's or department's obligation to provide a child with
accommodation under the Children Act 1989.

The *Homelessness Code of Guidance* (1991) makes reference to
the Children Act in pointing out that the government does not
wish to see a situation arising whereby young people are being sent
between social services departments and housing departments
with no one taking responsibility (6.16).

The experience to date of some voluntary sector agencies in the
North East suggests that most young people are not yet benefiting
from the introduction of the Children Act. Indeed, as feared,
some housing departments seem to be using the Children Act
as a means of passing any possible responsibility towards a homeless young person on to the social services department. One
example involved a 16-year-old homeless, pregnant woman
who approached her local housing department as homeless and
was told to seek assistance from Social Services as the introduction of the Children Act meant that she was now their
responsibility.

At this early stage it appears that, in some local authorities,
there are still no clear policies between Housing and Social
Services as to which young people are to be accommodated, and
who will accommodate them, and where. Social Services in some
areas appear to be willing to provide money for certain homeless
young people to stay in bed and breakfast accommodation.

It is thought that young people leaving care are now receiving
better treatment than before from some local authorities. Prior to
the introduction of the Children Act, research carried out by
Housing Aid for Youth into children's homes in south-east
Northumberland found that the majority of young people returned
to their parental home; it is thought that the main reason for this
was because there was no alternative accommodation for them.
Out of the 303 young people who took part in the survey, only 14
were housed by either local authorities or housing associations –
with equal numbers of male and female being housed.

In the absence of any priority status under the Housing Act 1985

Part III, there is a general reluctance on the part of both local authorities and housing associations to house under-18s. This is because a tenancy cannot be granted to a minor; however, it is possible for both local authorities and housing associations to enter into a contractual licence agreement with someone under 18, but some are only willing to do so if there is a rent guarantor.

The absence of a clear priority for young people leaves many in the group of the hidden homeless. Part of this group are those who sleep rough. The 1991 Census attempted a count of these rough sleepers and the results are interesting, both for their obvious under-counting of those involved and for the great gender divide. See Table 7.3.

Table 7.3   Number of rough sleepers by gender

Area	Men	Women
Inner London	906	166
Outer London	113	12
Metropolitan districts	264	32
English districts	962	197
Scottish authorities	122	21
Welsh districts	30	2
Total numbers	2397	430

Source: OPCS Census Monitor[1]

Anecdotal evidence from the Census collection time suggests that some local authorities 'cleaned up' their area beforehand, presumably for political reasons of their own, which throws doubt upon the accuracy of the figures and, of course, some rough sleepers may have avoided the count for reasons of privacy or fear of entanglement with authority. More recent discussions suggest that the Census has simply revealed the dual nature of rough sleeping; in London it is a highly visible phenomenon in open air city centre sites, while elsewhere there is more opportunity for people to sleep in small individual locations which reduces their chances of being counted (Birch, 1993). These factors may account for the low total and the fact that 315 authorities appeared to have no rough sleeping men and 383 no rough sleeping women. It remains true that rough sleeping is not a common route for women to take. This is hardly surprising, given the greater risks of sexual attack for women. This is compounded by the ideas informing a phrase such as 'a woman of the street' which is common parlance for a prostitute and conveys the view that women have no proper place in public. Mary Daly's report (1993) on homelessness in Europe comments that, while initial causes may vary across countries, male and female homelessness are decidedly different phenomena leading to a concept of two different routes to and forms of homelessness. Men's homelessness is more public in nature. It is precipitated by material changes and is more public in its manifestation. Women's homelessness typically arises from relationship problems and is 'solved' through private channels.

The more public face of men's homelessness drew government attention to the rough sleepers issue resulting in a number of specially funded programmes in the early 1990s.

One of these initiatives was the Housing Corporation's Rooflessness Package funded in June 1990 by central government in an attempt to provide hostel places and shared or self-contained accommodation for rough sleepers. Statistics from the Housing Corporation (Cheeseman, 1992) show interesting gender differences. Not surprisingly, only 26 per cent of those helped by the Rooflessness Package were women, which compares poorly with their take-up of 43 per cent of housing association tenancies in the same period. The Housing Corporation suggests that women may be seen as more vulnerable and therefore are likely to be rehoused more quickly, thus bypassing the Roofless Package altogether. Further research is needed to determine the truth of this. Another factor is that homeless women may be reluctant to accept help if it is offered by male volunteers. This is made more probable when set against the findings of the voluntary sector which document the numbers of young women leaving home because of physical and sexual abuse (Sexty, 1990).

While women make up a minority of those helped through this programme, 51 per cent of them were from Black and ethnic minority communities compared to only 19 per cent of male clients. A further difference is that the women were considerably younger than the men; 78 per cent were under 25 years old compared to only 43 per cent of male clients. Why young Black women should be particularly vulnerable to homelessness is an area which needs to be further researched.

Where women are given permanent accommodation difficulties

can arise with the allocation of properties. For example, in Northumberland, two of HAY's clients were each housed in bedsit blocks, with six bedsits to each block. They were the only women in each of the blocks and suffered from extreme harassment. It proved very difficult for them to be rehoused because of policies which looked primarily at housing need in terms of measurable qualities, e.g. security of tenure, over-crowding, the physical quality of the accommodation, etc.

Finally, there is a need to be aware of the biggest group of hidden homeless, those who rotate around their friends, sleeping on a floor here and a sofa there. Research undertaken in Glasgow by the Scottish Council for Single Homelessness (1989) revealed that 23,000 people or 62 per cent of the waiting list were staying care of another address. Women made up 5000 or 21 per cent of these hidden homeless.

The ability to sustain accommodation and its affordability are closely entwined. The next section examines the changes in housing policy, social security and employment policy which have militated against young people seeking independent living.

CHANGES IN SOCIAL SECURITY AND EMPLOYMENT POLICIES

The first major change was the Board and Lodging Regulations 1985. This Act was targeted at people under 26 living in bed and breakfast accommodation. The majority were not 'living the life of Riley' on the dole but had taken up Norman Tebbit's suggestion that they get 'on their bike' and look for work. Research from the period shows young people moving away from areas of deepening unemployment in an effort to get a job. The changes brought about by the 1985 Act cut across these efforts: under the Act those under 26 were eligible for the householder rate of board and lodging allowance for a limited period only.

These limits were set at eight weeks for London and the big cities, four weeks for most districts and two weeks in seaside districts. At the end of this period the young person was reduced to the considerably lower non-householder rate of benefit, leaving the shortfall to be made up from Supplementary Benefit intended for food and other expenses. This had the effect of making board and lodgings unaffordable for young job-seekers and other young people unable to find any other accommodation.

The Social Security Act 1986 made major changes to the benefit system and to the financial standing of unemployed young people. The first alteration was the abolition of the householder and non-householder distinction which had recognised the greater financial demands placed upon a young person by independent living. In 1985 half of those aged 21–24 who were Supplementary Benefit claimants were claiming householder rate; all of these suffered financially as a result (Thornton, 1990). The ideological thrust was that young people should stay at home until they had the financial means to justify a move to independence. This was made more clear in the introduction of age banding which created three classes of young people: the under-18s, the 18–24 group and those aged 25 and over. These three groups were paid different levels of the newly introduced Income Support which replaced Supplementary Benefit. This was more than a relabelling of an established benefit. Income Support is payable for two weeks in arrears, unlike Supplementary Benefit. The Board and Lodging Regulations were abolished under the 1986 Act, but this was not an easing of the problems of the residents of such places for, from April 1989, the lodging element was met by Housing Benefit and the board element from Income Support. Because Income Support is paid in arrears a claimant may find his or her landlord/landlady unwilling to wait for the money owed. This is compounded by the well-known slow processing of Housing Benefit claims. Those operating board and lodging establishments are running a business, or certainly not a charity, and are unlikely to be able to wait for what can be many months. Moreover, when payments are processed they are unlikely to cover the charges. Housing Benefit ceilings are based on rent levels in a district, not on hotel charges. Similarly, the element of Supplementary Benefit which was paid to cover the extra expenses faced by those with no cooking or laundry facilities and therefore obliged to use take-aways and launderettes has been dropped, leaving all these charges to be paid from Income Support and, for young people, a reduced Income Support based on their age not their needs. Board and lodgings are no longer an affordable route to independence.

In July 1989 the age barrier between the under 18 year olds and those in the band 18–24 was lifted making the very young eligible for the same level of benefits as the older group. However, the assistance offered by this policy U-turn had already been wiped out by the 1988 Social Security Act which had lifted the eligibility threshold for Income Support from 16 to 18, making very few young people (except those with disabilities or those with a child) able to make a claim. The raising of the threshold was timed to coincide with the introduction of the two-year YTS programme with its 'guaranteed' place for each school leaver. The effect of these two policies was to make YTS compulsory and, for those unable to gain a place, no entitlement to income other than a time-limited barest transitional relief.

For those young people on a bridging allowance or a training scheme, no assistance on any front will be forthcoming from the DSS, unless they are for some unusual reason also receiving Income Support. The government considered the private rented sector crucial to its housing policy, yet landlords in this sector invariably deter tenants by asking for deposits and rent in advance for accommodation. The same is often true of bed and breakfast landlords. Deposits are not available from the DSS for anyone; rent in advance is payable under discretionary Income Support. The young person on a training scheme or bridging allowance can therefore forget about receiving any assistance from this source, and can usually forget about private renting as an option. Housing Benefit regulations can present further difficulties for those hoping to obtain private rented accommodation. Where a person is con-sidered to be 'over accommodated' or is paying a rent considered by the local authority to be above the market value for properties of that type and in that area, a restriction on the amount of benefit payable to the young person may be imposed. Therefore, even if a young person manages to find rented accommodation, with more than one bedroom he or she may be considered to be over accom-modated and again there will be a restriction on Housing Benefit.

If a young person does succeed in getting a tenancy, the next problem will be in obtaining furnishings. Nearly all local authority and housing association properties are let unfurnished.

The 1988 Act also abolished the single payment which had often been used to help a young person acquire the basics of a second-hand cooker, a bed and chairs. The abolition of this single pay-ment leaves the Social Fund as one of the few sources of help.

The workings of the Social Fund are such that many under-18s who look to it as a source of help will inevitably be disappointed. Those who are not in receipt of Income Support cannot usually receive any assistance from the Fund. For those on bridging allowance or a training scheme, no help will be available from the

DSS. Those on Income Support may get a grant from the DSS if they fall into a very restricted set of categories, e.g. leaving care, coming out of some institution, etc. But the majority of young people who are on Income Support must expect to receive a repayable loan as their only form of assistance. To get this they will have to have been in receipt of Income Support for 26 weeks or more and, given that Income Support is only payable to 16 and 17 year olds in restrictive circumstances and often for short periods of time, the likelihood is that those few 16 and 17 year olds who receive Income Support will not have been receiving it for a sufficient time, and are therefore ineligible for the budgeting loan. In short, they are unable to get any state help in obtaining furniture.

All this applies equally to young men and young women. There is, however, a group of young women who appear to be made incredibly vulnerable by these benefit rules – pregnant young women. As with all under-18s, pregnant young women are usually expected to work or be on a training scheme. They are allowed to receive Income Support at the age-related amount once they are within 11 weeks of having the baby. This, therefore, means that until the young woman receives Income Support she will not receive all the benefits automatically received with Income Support; for example, milk tokens. She cannot usually make any application to the Social Fund before receiving Income Support. She still may not succeed in obtaining any help from the Fund (beyond her entitlement to a maternity grant) as she will not have been receiving Income Support for the required 26-week period, and therefore cannot be considered for a loan.

Problems inevitably arise for these young women when they try to prepare for the birth of the baby. If they are setting up home for the first time, they have to obtain furnishings as well as bedding and clothing for the baby. It is impossible to achieve this on a YTS or equivalent wage and, if no help is forthcoming from the Social Fund, the woman may find herself without any furnishings or items for the baby, even though she is expecting in 11 weeks or less. These particular problems are experienced only by those under 18 – arguably the most vulnerable of all young women. Women aged 18 or over can receive Income Support as of a right, though those aged 18–24 will receive less than those aged 25 or over. The issues here are not only about affordability and sustaining a tenancy, but also about health – both the young woman's health and that of her baby. This is best illustrated by the example of 'Kim', one of HAY's clients.

Kim was 17 years old and pregnant when she came to HAY. She had been told to leave the parental home because of a relationship breakdown with her parents. Kim was living in with friends, but on a temporary basis.

The housing problem was straightforward enough to resolve in terms of Kim being accepted as a priority case by the local authority. But the benefit problems were much greater. Kim had no income at all when she came to HAY. She had been relying on her friends to feed her since leaving the parental home. Because she was under 18 and only a few months pregnant, Kim had no general entitlement to Income Support. Fortunately, Kim was seeing her doctor about a stress-related illness and she could therefore be advised that she should receive Income Support on the grounds of ill health, provided her doctor would confirm her unfitness for work. However, it was apparent that her doctor was unlikely to consider Kim unfit for work for the length of her pregnancy, and it was necessary to advise her that, until she was within 11 weeks of having the baby, she would have to take up a training scheme if she was to have any income.

There are several problems with this. First, the difficulty of finding an employer willing to take on someone who will have to go on maternity leave in three to four months. Second, as Kim would no longer be entitled to Income Support on taking up a training scheme she would lose her entitlement to milk tokens, which are worth about seven pints of milk a week. She would also be unable to look to the DSS for any help with furnishings when she was rehoused.

It is interesting to look at how someone like Kim can, in fact, afford to keep a tenancy and prepare for her baby's birth.

Table 7.4 presents a crude budget that was worked out with one of HAY's clients who was on a Youth Training Scheme. This budget was done before the abolition of the Community Charge and its replacement with the Council Tax, which under the new system Kim would be exempt from paying. However, the impending introduction of VAT on domestic fuel and the raising of water and sewerage costs will easily wipe out the small saving that she makes.

When the baby is born, Kim will receive less than a mother aged

Table 7.4 A weekly household budget for a young person on a YTS

Item	Weekly amount
Rent	£ 2.50
Water rates	£ 2.50
Community charge	£ 2.00
Housekeeping	£15.00
Gas	£ 4.00
Electricity	£ 4.00
Clothing	£ 2.00
TV licence	£ 1.00
Transport	?
Entertainment	?
Cigarettes	?
Total	£33.00

25. Her income will be £62.20 compared to £71.05 for someone aged 25.

Had Kim been in this situation before the benefit changes in 1988, she would have received exactly the same income as any other householder, and age would not have been a relevant factor – either before or after the baby was born.

The real problem with so many young women in Kim's situation, even if they are not pregnant, is that they have no choice but to obtain a tenancy somehow and then sustain that tenancy. Failure to do so simply results in homelessness.

Sustaining a tenancy is becoming more difficult for young people as unemployment for their age group worsens, and the number of YTS trainees who find a job at the end of their scheme remains half those who complete courses (OPCS, 1993). See Table 7.5.

Table 7.5 Unemployment rate by age and gender in 1992

Age	Men %	Women %
16–19	18.7	13.84
20–29	15.3	9.4

Source: OPCS, 1993

Table 7.6 Average gross weekly pay for full-time workers

Age	Men	Women
< 18	£110	£107
18–20	£163	£146
21–24	£229	£189

Source: New Earnings Survey, 1991

Table 7.7 The Housing Benefit poverty trap 1994/5: combined tax and benefit rates and disposable income cumulative reductions for each £1 of gross earnings

	Single person or couple
Income tax @ 20 per cent	20p
National Insurance @ 10 per cent	10p
Net earnings	70p
Housing Benefit @ 65 per cent	46p
Council Tax benefit @ 20 per cent	14p
Net disposable income	10p

Source: Wilcox, 1993

For those in work their wages remain low, a position made worse by the abolition of the Low Pay Council. Evidence from the New Earnings Survey 1991 indicates the poor wages paid to young people and in particular how young women quickly fall behind their male counterparts. See Table 7.6.

For those in work the impact of Housing Benefit tapers means that young people are left with only pennies from each pound they earn over the benefit threshold. See Table 7.7.

This compares to the same person keeping 13 pence in 1992/3 and 33 pence ten years ago. This will be examined in greater detail later in the chapter when we consider affordability and housing association tenancies. Taken together, we have a situation where many young people cannot find work and are kept in poverty by social security policies which militate against them and which, when they do find work, keep them in a poverty trap through a combination of poor wages and welfare benefits which taper off severely.

The changes in benefits have impacted particularly on young people but the changes in housing have also played a part in reducing the options of this group.

HOUSING CHANGES

In 1988 the Housing Act attempted to give the kiss of life to the moribund private rented sector, but evidence from an OPCS study carried out soon after the introduction of the legislation indicated that, while landlords might have viewed the changes as a positive step to making renting out more viable, the impact on tenants was detrimental (*Housing*, October 1991). The London Research Centre has since found that 50,000 tenants have experienced attempts to evict them. In 1989 an investigation by Shelter in Preston and Sheffield revealed a 50 per cent increase in the number of cases of harassment and illegal evictions. Ironically, there has been an increase in private renting but this has developed out of the housing bust. Properties rented out because their owners cannot sell are likely to be shorthold tenancies with rent levels linked to the monthly mortgage repayment level; in short, they offer little relief to a low paid or unwaged young person, should he or she be seen as suitable.

The same Act changed housing association capital finance and the rent regime of this sector. The move to mixed funding and the decrease of housing association grant (HAG) (with further decreases promised) have led inevitably to higher rents. The National Federation of Housing Associations's *CORE Quarterly Bulletin* (January–March 1992) comments on the decrease in the proportion of new tenants who are in employment from about 30 per cent two years previously to about 25 per cent in the first quarter of 1992. The reason is plain: rises in rent, coupled with Housing Benefit tapers, create a poverty trap attracting either the well paid or the unwaged and keeping out the low waged. See Table 7.8.

The same bulletin reveals rents for a one bedroom property ranging from £19.73 for fair renting in Merseyside to £32.66 in the South East, and for assured renting £26.16 in Merseyside to £39.80 in the South East. What these figures mean for young people is best explored by looking at the finer grain statistics provided by

Table 7.8   Affordability rates by region and tenancy type

	Fair relets %	Assured lets %	Assured relets %	All %
East	20.2	26.9	26.2	26.3
East Midlands	22.9	30.0	27.0	27.7
London	21.1	31.6	27.4	27.3
Mersey	13.5	27.2	21.0	21.1
North	20.2	26.2	24.0	24.4
North West	16.1	32.3	26.6	27.6
South East	22.1	30.5	27.7	28.6
West	22.7	28.5	28.2	28.1
West Midlands	21.3	28.2	26.7	26.7
Yorkshire/Humberside	21.5	28.5	26.5	27.2
England	20.4	29.0	26.4	26.8

Source: NFHA *CORE Quarterly Bulletin*, July–September, 1992

Table 7.9   Rent affordability in the North East

Earnings	Housing benefit	Tenant pays
A 24 year old in 1991 paying rent of £32.50 for a new build one bedroom flat		
£60 per week	£15.47	£17.03 – 28% of net income
£80 per week	£ 2.47	£30.03 – 38% of net income
Tenant gets a new job on her 25th birthday		
£110 per week	£0	£32.50 – 30% of net income
A 24 year old in 1992 (projected picture from 1991), paying rent of £48.96 for new built one bedroom flat		
£60 per week	£31.93	£17.03 – 28% of net income
£80 per week	£18.93	£30.03 – 38% of net income
Tenant gets new job on her 25th birthday		
£110 per week	£4.56	£44.40 – 40% of net income

Source: NFHA, *Northern Region Annual Report*, 1991

the regions, in particular the North where wage levels for men and women are the lowest in Britain. See Table 7.9.

As the tenant approaches the income level of the lowest paid quarter of the workforce in the region, her rent is twice the NFHA affordability level. While single people have always been a successful group in terms of their access to housing association properties (16 per cent of new tenants in the quarter of January–March 1992, NFHA *CORE Quarterly Bulletin*) it must be a matter of concern that rents are now keeping those young people in poverty or leading them to reject employment offers.

Affordability levels are not just issues for housing associations. While average renting costs for the private rented sector are more difficult to pin down, there is widespread variation in the rents charged by local authorities. See Table 7.10.

Table 7.10   Average weekly unrebated rent for a one bedroom flat

	Lowest	Highest
Metropolitan districts	£16.83	£28.79
Districts	£14.90	£41.90
Inner London	£28.31	£45.95
Outer London	£15.58	£58.70

Source: CIPFA, *Housing Rent Statistics*, April 1992

HOUSING AID FOR YOUTH

Housing Aid for Youth (HAY) is a small voluntary project in Blyth, south-east Northumberland which offers advice and assistance to men and women aged 16–26 on housing and benefit matters. HAY also has six units of accommodation for young people leaving care.

From the few regional studies on women's homelessness, Shelter concludes that young women are most at risk. They conclude this from the heavy over-representation of women aged under 26 among the homeless. A survey in Bristol (Bristol City Council, 1988) revealed that women aged 16–25 represent less than a third of the general population of single women, but nearly three-quarters of single homeless women. At a local level Cumberland House (a Newcastle-based project providing supported accommodation for women aged 17 and over) found that in 1991 27 of the 34 women assisted were in the 17–25 age group. More worryingly, 22 of these women were under 21 years old. Nearly half of HAY's clients are women (49 per cent) and over half of this group are either 16 or 17 years old. This again is the experience of Stepping Stones (a Newcastle project providing accommodation for young men and women). The majority of those contacting Stepping Stones are women and among these the under-18s are over-represented – 258 of the 388 women in contact with the project.

There seems to be a consensus of opinion among those organisations in contact with young women that their homelessness is of the hidden variety, that is they are sleeping on friends' floors, relations' settees or putting up with relationships they no longer want. At HAY the majority of those clients who have been sleeping rough are male; a very small percentage of female clients have ever slept rough. This also seems to be the experience of other projects, such as Stepping Stones. However, the number of young women approaching HAY who might be termed 'potentially' homeless, that is, those who are likely to be homeless in the near future, is higher than that for men, amounting to 56 per cent of all women clients as opposed to 46 per cent of all men.

It is Shelter's view that family problems are a major reason for young people leaving the parental home. This term is used to cover situations where young women have been subjected to emotional deprivation or sexual or physical abuse. This again accords with HAY's and Stepping Stones' experiences. Most of the young women approaching HAY explain their homelessness (or threatened homelessness) in terms of irretrievable relationship breakdown with parent or parents. In addition to this, well over 10 per cent of women clients state that they have experienced violence or been abused. This figure cannot represent the true number of women experiencing abuse as it is derived from information provided by clients in the first few interviews. Many clients will never discuss abusive situations and some will only do so after a long period of contact with the project. Figures collected by Shelter from Newcastle City Council showed that more than 90 per cent of the teenagers whom they accepted as homeless had to leave because they had been evicted or because of violence or sexual abuse in the family (*Roof*, 1985). In general, it appears that most young women leave the parental home (if they have one) because they find that environment intolerable. Those leaving care obviously have no parental home to leave.

Cases such as Kim's (quoted earlier) illustrate the frustration faced by those working to alleviate young women's homelessness and poverty. So many of the problems outlined in her story arise directly from central government policy. Nevertheless, it is important to keep in mind the issues that can be addressed.

POLICY INITIATIVES

In spite of the verbal enthusiasm given to the idea that young people themselves ought to be involved in policy-making, or at the very least consulted, the reality is that it still does not seem to happen. One of the reasons may be that the forums have not existed where young people can comfortably give their views. This problem is now being addressed by organisations such as HAY and a Newcastle-based group called Kids Moving On (KMO). KMO has a member of staff who concentrates on working with young people on the issues of employment and accommodation. HAY is currently undertaking training with some young people and through the direct involvement of other young people is preparing material to help them take a more participative role. The Newcastle Young Women's Project has already got together a group of young women, who attended a Shelter conference to give their points of view on women's housing issues. The KMO group hopes that a north-east young persons' forum can be formed, to provide a means of gaining their perspective on housing and benefit issues. It is also hoped that there may be scope for young women either to have their own forum or at least be able to articulate their own particular problems and needs. Once such a group emerges, it will then be up to housing providers and others to take it seriously and put their ideas into policy.

At the same time a group has recently been set up as an initiative of the Single Homeless on Tyneside (SHOT) and Northumbria Probation Service which is looking at the services for homeless women and the practices of organisations working with these women.

SHORT-TERM HOUSING OPTIONS

In hostels which provide accommodation for both men and women, there can be a reluctance on the part of women to enter a mixed establishment, particularly where they are fleeing abuse from a man. In other mixed hostels the demands of men may overwhelm the whole project, leaving no places for women. It is important that a certain number of beds are kept for women and, of course, that women have a choice of mixed or single sex hostels. This is more vital as it becomes apparent that bed and breakfast is not usually an option for young women. This is either because of a lack of bed and breakfast places willing to accept young women, or again because they are occupied more or less exclusively by men. It may, therefore, be necessary to consider alternatives to bed and breakfast. One such alternative is the Lodging Scheme which is run for men and women. The scheme was set up by HAY in response to people coming forward who were interested in taking a young lodger. HAY visits these prospective landlords and land-ladies, examines the facilities and discusses the rent. The scheme has run successfully for more than three years, allowing a number of young women to make a safe transition from leaving the paren-tal home or care to their own tenancy. A local information and campaign group conducted some research into what sort of accom-modation was needed in the region in the opinion of the organisa-tions already working in the field. From this it was determined that more good quality temporary accommodation was needed, in particular for young women fleeing abuse (perhaps some form of safe house) and for young lesbians.

Two new local temporary accommodation options have become available for young women in the last year. One in North Tyneside was set up by the local authority, and provides accommodation for any homeless woman aged 16 or over. The second is a project set up in Newcastle specifically for pregnant young women (aged 16–25) and those with children.

LONG-TERM HOUSING OPTIONS

Newcastle City Council continues to be the only authority in the North East that will accept 16 and 17 year olds as a priority under the Housing Act 1985, Part III on the grounds that their age alone makes them vulnerable under the terms of the Act.

While accepting young people the local authority has had a corporate concern for the problems faced by young people trying to live independently. In response to this an Officer Working Group composed of representatives from Social Services, Housing and the voluntary sector was formed in 1990 to prepare proposals for implementing supported accommodation throughout Newcastle.

The aims of the project (named First Move) have been to divert young people from homelessness and failed tenancies, and offer an integrated service to support young people moving into independence. The key elements of the proposal are:

- integrated service provision able to respond to the multi-faceted needs of young people;
- a corporately resourced service;
- a service which recognises the need for individual and group support;
- a service which addresses the need for user participation (Newcastle MBC, 1991).

In the 1992/3 financial year a scheme of 15 furnished flats has been let, and it is hoped that these will be the first of a number of supported dwellings for young people providing a mix of accommodation throughout the city.

The First Move initiative is a further step in the city's development of furnished accommodation, bringing together this initiative and the long established (1983) Single Person Support Workers. These workers originally worked from a central office base but now operate on an outreach basis giving budgeting advice, advice on welfare benefits and general support to young people. The furnished lettings strategy has been operating since the early 1990s when a small number of tower blocks, previously considered difficult to let, were refurbished providing furnished flats with 24-hour security. Evidence from those working with young people underlined the considerable benefit to both young tenants and the housing department in providing either furniture or furnished lettings. It was suggested that the provision of furniture reduced the turnover of flats by approximately 30 per cent, reflecting both a saving to the authority and an increase in the ability of young tenants to make a success of their tenancy. This led the authority to develop the range of furnished dwellings. The pilot scheme, begun in April 1992 (Newcastle MBC, 1992), provides part-furnished accommodation for those otherwise unable to provide furniture for themselves. Basic furniture is bought in as new or provided from stock refurbished by the Community Furniture Service. This initiative has been operating since 1987, supported initially by Inner City Partnership. The Furniture Scheme delivers furniture by appointment and provides a guarantee repair/ replacement service within five working days for all items rented by the tenant subject, of course, to an assessment of whether damage has been due to wear and tear or wilfully inflicted. It is anticipated that this part-furnished initiative will pay the same dividends as the fully furnished flats but an evaluation has yet to be undertaken. Problems of security in dwellings without 24-hour surveillance are, of course, a possible issue, but dispersed furnished schemes in cities such as Glasgow have demonstrated that losses are not necessarily large scale.

CONCLUSIONS

There are still too many young women like Kim, made homeless by changes in personal circumstances but then rendered destitute by aggressive housing and social security rules. In the long term, there is an urgent need for young people to have benefit fully restored to them at a rate which reflects living costs, and for those under 18 to have a right to Income Support. In particular, the policies which militate against young pregnant women should be urgently reviewed.

In areas where there is a shortage of appropriate sized accommodation there should be a relaxation of the Housing Benefit over-accommodation ruling. There should be a review of the Housing Act 1985, Part III to extend the official definitions of homelessness and priority need to ensure that young people and those escaping harassment and abuse on the grounds of their sex or sexuality (and race) are covered by the legislation. In order to make this a reality and not simply a further burden to local authorities, there must be a better financial commitment to social housing. Whether this comes from local authorities or from housing associations is not relevant, though all providers need to take responsibility for housing the homeless. The needs of particular women, such as Black women, women with disabilities, the very young and lesbians, should be planned for, particularly in respect of short-term accommodation, though it is also necessary for such women to have freedom of choice in respect of their longer-term housing options.

NOTE

1 Thanks to Dr Dan Dorling, Joseph Rowntree Fellow at the University of Newcastle for production of census tables.

REFERENCES

Birch, J. (1993) 'Now you see them, now they don't', *Roof*, May/June: 32–5.
Brailey, M. (1985) *Women's Access to Council Housing*, Occasional Paper No. 25, Glasgow: The Planning Exchange.
Bristol City Council (1988) *Single Women and Homelessness in Bristol*.
Central Statistical Office (1993) *Social Trends 23*, London: HMSO.
Cheeseman, D. (1992) 'Monitoring the Rooflessness Package in London, February–September 1991', *Homelessness Statistics*, Seminar Papers from Statistics Users Council Seminar, 16 December 1991.
Children Act Housing Action Group (1992) Briefing Paper No. 1, London: CHAR.
CIPFA (1993), *Housing Rent Statistics*, April 1992.
Daly, M. (1993) *Abandoned: Profile of Europe's Homeless People*. The second report of the European Observatory on Homelessness, Brussels: Fédération Européen d'associations nationales travaillant avec les sans–abri.
Department of the Environment and Department of Health, Welsh Office (1991) *Homelessness Code of Guidance for Local Authorities*, third edition, London: HMSO.
Dibblin, J. (1991) *Wherever I Lay My Hat – Young Women and Homelessness*, London: Shelter.
Greve, J. and Currie, E. (1990) *Homelessness in Britain*, York: Joseph Rowntree Foundation.
Housing (1991) 'Private sector shows little sign of life', October, monthly Factfile series, p. 23.
Kay, H. (1992) *Guide to Means Tested Benefits for Single People without a Permanent Home*, 13th edition, London: CHAR.
National Federation of Housing Associations (1991), *Northern Region Annual Report*, Newcastle upon Tyne: NHFA.
National Federation of Housing Associations (1992) *CORE Quarterly Bulletins*, London: NFHA.
New Earnings Survey (1991) London: HMSO.
Newcastle MBC (1991) *Review of Housing Allocations Policy: Progress Report*, Housing Committee, September.
Newcastle MBC (1992) *Development of Furnished Accommodation: Progress Report*, Housing Committee, October.
Niner, P. (1989) *Homelessness in Nine Local Authorities: Case Studies of Policy and Practice*, London: HMSO.

Office of Population and Census Statistics (1993) Census 1991 (Crown copyright).

Scottish Council for Single Homelessness (1989) *Women's Homelessness: The Hidden Problem*, Glasgow: SCCH.

Sexty, C. (1990) *Women Losing Out: Access to Housing in Britain Today*, London: Shelter.

Thornton, R. (1990) *The New Homeless*, London: SHAC.

Wilcox, S. (1993) *Inside Housing*, 19 March.

Women's National Commission (1983) *Report on Homelessness amongst Women*, London: Cabinet Office.

Chapter 8

'The struggle has never been simply about bricks and mortar'[1]
Lesbians' experience of housing

Julia Smailes

Lesbians are practically invisible in all discussions of housing policy. Yet lesbians know that housing can be a big issue in their lives, and that they have limited options if they want to be open about their sexuality and near other 'out' lesbians for support. There has been little research into lesbians' housing needs and choices, and such research as there is has been done on a small scale by committed groups of lesbians.

This chapter examines lesbian issues in housing, rather than lesbian and gay issues, and the experience of being a lesbian is placed within the general oppression experienced by women in a patriarchal society. Some of the issues are common to both lesbians and gay men; however, men do not experience sexism or the issues faced by women living a life without reference to male approval. Heterosexism, defined by the Labour Research Department as 'the system of beliefs, attitudes and institutional arrangements which reinforces the view that everyone is, or should be heterosexual' (Labour Research Department, 1992: 4), leads to the invisibility of lesbians. Although this invisibility and denial of lesbian existence causes many problems for women questioning their sexuality, it does not protect lesbians from harassment and discrimination.

The State, religious organisations and society in general do not recognise lesbian and gay relationships at all. As well as the personal suffering this can cause, it also means that lesbians do not have any property or pension rights on the death of a partner and are often rejected by their partner's biological family.

There is no protective legislation for lesbians when faced with discrimination in British society. In fact, the current trend in legislation is to encourage discrimination without being explicit

about it. Clause 28 of the 1?
it illegal for local authorities
sexuality'. It is all right to be
anybody else to be one.

This chapter looks at the way
against lesbians affect housing opt
two surveys carried out in June
association policies towards lesbian.
ence of lesbians living in a town in Y

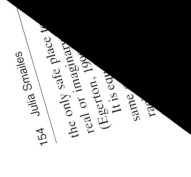

HOUSING ISSUES FOR LESBIANS

The way housing is provided influences   ⌐ut where and
with whom we live; for example, incre ⌐ingly, two incomes are
needed to service a mortgage and social housing providers decide
what constitutes a legitimate household.

Housing is for heterosexuals. There is an assumption of hetero-
sexuality throughout housing policy. 'Families', a man, a woman
and child(ren), are the top priority with everything else being
some form of special need, including single parent families, i.e.
families with a man missing (90 per cent of single parents are
women).

Within this lesbians are either invisible or seen as the problem,
rather than having a problem. Sexuality may not be an issue to
most housing providers but housing is certainly an issue for lesbians: 'The London Lesbian and Gay Switchboard, moreover,
receives more phone calls in connection with housing than any
other telephone agency, including housing specialists' (Dibblin,
1988: 25). There has, however, been little research into housing
need among lesbians – certainly no major publicly funded
research.

It cannot simply be assumed that research into women's housing
issues, done from a (covertly) heterosexual perspective, necessarily covers lesbian issues. An example of this is the meaning of
'home' for women. Feminists rightly challenge the notion of a
home as a haven from the world as being a male view and not
universally applicable. 'Home' and 'work' are not separate places
for heterosexual women who still do almost all the housework and
childcare. 'Home' is also a place where women experience
violence and abuse from partners. However, for lesbians, who are
not at liberty to live as lesbians in the outside world, home can be

be a lesbian. 'Home, for all lesbians, is the
place where we feel safe, loved and validated'
0: 76).
ally important to recognise that not all lesbians share the
experience and that other forms of discrimination, such as
ism, ageism and discrimination against disabled people, combine and give rise to different needs.

In outlining some of the housing problems facing lesbians it has
been necessary to draw extensively on the few reports and articles
that are available.

Disabled lesbians

All disabled people have difficulty finding accommodation suitable
to their needs and will continue to do so as long as 'general needs'
are so narrowly defined as to exclude them. Disabled lesbians face
extra problems, however. It is difficult for disabled people in
residential institutions to assert any sexuality, but to come out as
lesbians leaves them vulnerable: 'If you come out, will you still get
care at all or get it sympathetically? . . . (P)eople know everything
about you . . . (Y)ou can be refused support and co-operation in
areas of which people disapprove' (Cookson, 1989: 8).

This also applies to disabled lesbians living independently but
with support. There is rarely a choice of carers:

> There is no guarantee that they won't be racist or homophobic
> and the carer has immense power as they can withdraw support
> . . . if the support is from friends or family, there is a need to
> keep up a good relationship with them and this often means not
> discussing sexuality.
>
> (Cookson, 1989: 8).

Agencies supposedly facilitating independent living may not
acknowledge lesbian relationships, as in the case of Joan in the
Lesbian Information Service study who 'had to move from
partner's home because of access difficulties. Disablement Officer
could have told us about certain grants and benefits enabling
adjustments to property' (Lesbian Information Service, 1988: 31).
She was allocated an adapted flat as a single person. Imagine the
officer being responsible for separating a married couple by failing
to give them the information they needed!

Older lesbians

Most old people do not go into sheltered housing or residential care. However, for those who do, they face an ageist assumption that they are no longer sexually active and a heterosexist assumption that their past has been heterosexual. Older lesbians are thus denied their history. In one residential care home which was visited there is a wedding photo display board in the entrance lobby with pictures from the 1920s to today, celebrating the heterosexual ritual and excluding lesbian experience.

Older lesbians are afraid to come out because of the fear of victimisation or withdrawal of support. Those who do come out may experience abuse from wardens and management.

Moving in order to receive physical support may also mean isolation and the loss of a supportive lesbian network. The following example is taken from a report by the Lesbian Workers Group at Pensioners Link (Aimsworth, undated).

Lily and Nora lived in the same residential care home and both were very dependent on staff for physical help. They experienced abuse for showing affection to each other. On the night Nora was dying Lily was not allowed to be with Nora and was forced to go to a social event. 'Lily was told to stop whining [as] Nora was going to be fine . . . [then] . . . finally a care assistant told Lily that Nora had died. Lily died a week later'.

Lesbian networks themselves are dominated by younger women and older lesbians, who may have always led a more isolated life, and may have less access to the support which could help them to ask for what they want. The report says of an older lesbians group, 'the housing needs of lesbians have proved to be an issue of great concern to the group'.

Black and Irish lesbians

Black people experience discrimination in both access to and quality of housing in Britain. The experience of racism means that family and community play an important role for Black people, and Black lesbians can be faced with risking the loss of that support by coming out without finding equivalent support in a White-dominated lesbian community. 'Single black women might find it harder [to obtain housing] than a white lesbian, who at least has the option to hide her sexuality to avoid discrimination' (Anlin, 1989: 36).

Irish lesbians also face anti-Irish racism in Britain.

> Being a lesbian is tied in with my Irish nationality. . . . As lesbians we all have different experiences, and white middle-class British lesbians have privileges. . . . I would like to live with other Irish lesbians. Irish and black lesbians need other Irish and black lesbians in the house for mutual support.
>
> (Anlin, 1989: 25)

HOMELESSNESS

Homelessness has been defined as a continuum ranging from sleeping rough through hostels to various insecure and temporary arrangements, with women's homelessness often concealed at this latter end (Watson and Austerberry, 1986). With the rise in 'statutory' homelessness it is easy to view other homeless people as 'not as homeless' and therefore put off dealing with their problems. Here homelessness will be considered in relation to young lesbians, lesbians leaving a matrimonial home and lesbians experiencing harassment. Homelessness following bereavement will be discussed with succession rights.

Young lesbians are often thrown out by their parents or guardians for coming out, or leave because they are unable to come out: 'According to a survey of nearly 500 young lesbians and gay men living in London, carried out in 1983, more than one in ten had been thrown out of home because of their sexuality' (Foster, 1988: 20). Sandra Anlin's study of lesbians' housing experience for Homeless Action found that eight out of 20 homeless lesbians were: 'thrown out by parents for being a lesbian, or left home for same reason as unable to "come out" to them' (Anlin, 1989: 4). This happens at a time when young lesbians are already vulnerable, coming to terms with their sexuality in a heterosexist society. Local authorities, however, will only provide rehousing if they are 'vulnerable' under the 1985 Housing Act and few authorities recognise the vulnerability of (young) lesbians.

Many lesbians have been in heterosexual marriages before coming out. Sometimes they are able to continue living in the matrimonial home while sorting out an alternative; however, others leave to face homelessness or insecure accommodation: 'I left marital home. Had no money, no job and nowhere to go. Left behind everything I had. Had no choice' (Lesbian Information Service, 1988: 30).

The Lesbian Information Service (LIS) report includes a case study of a woman which highlights the chain of insecure accommodation in poor housing conditions she faced following her decision to leave her partner and children. She says: 'The reality of having left the (relative) security of heterosexual living finally hit me when I moved into a bedsit' (LIS, 1988: 37). For lesbians in this situation there is also a risk of losing custody of their children, just because they are lesbians. The woman in the case study 'failed to achieve even joint custody of my two children, giving up after a three year struggle' (LIS, 1988: 37). Ironically, had her children been with her at the outset the local authority might have had a duty to rehouse her. Other lesbians receive custody orders with conditions such as not letting a lover stay overnight.

Landlords and neighbours frequently harass lesbian residents: '45 per cent of those applying to Stonewall [a hostel for young lesbians and gay men] gave this as a factor in their homelessness and Homeless Action . . . agrees that it is *the* major cause of homelessness for lesbians' (Dibblin, 1988: 26). The following are examples drawn from the research:

> another woman was thrown out by a landlord who'd seen a lesbian poster in her flat.
>
> (Dibblin, 1988: 26)

> one lesbian couple had excreta put through the letterbox of their council flat, had had break-ins, and were subsequently forced to sleep with a knife by their bedside.
>
> (Anlin, 1989: 5)

> I was living with a friend of the family. He knew I was a lesbian but he kept trying it on with me. He gave me an ultimatum, either I went to bed with him or I left. I left.
>
> (Dibblin, 1991: 30)

The 1988 Housing Act was supposed to provide stronger protection against illegal eviction. However 'the increase in protection from harassment under the new Housing Act does not extend to harassment for being lesbian or gay' (Cookson, 1989: 10).

Hostels are a major source of emergency accommodation for single homeless people. However, the attitude that more men than women are homeless is reflected in the fact that there are nine bedspaces for men for every one bedspace for a woman (Dibblin, 1991: 35). However, hostels that do cater for women can be unsuitable for lesbians because of harassment. Many lesbians look elsewhere:

> The high incidence of intimidation directed at young lesbians and gay men in much of the emergency accommodation available for the young, single homeless is sufficiently well known to discourage them from entering such accommodation, forcing them to seek other, less satisfactory alternatives.
>
> (Foster, 1988: 20)

The LIS report details the difficulties that lesbians face in hostels, including a case study of a lesbian living in a hostel: 'Every time someone new comes I have to come out again. It's such a strain . . . it is bad for me if I don't tell them and it is bad for me if I do tell them' (LIS, 1988: 40).

Homeless Action operates a 50 per cent quota of lesbian residents and some of its houses are lesbian only. Twelve out of 14 respondents in the LIS study said that they would have welcomed lesbian-only accommodation at times of crisis in their lives. One problem with identifiably lesbian-only accommodation is vulnerability to attack: 'I don't think lesbian hostels are a good idea – it'll be a prime target to put a window out or spray paint the walls. They need more women's hostels with lesbian workers' (Dibblin, 1991: 53). At the moment, however, lesbians rarely have a choice about whether to live in lesbian-only hostels, particularly outside London.

The quotation from Dibblin also raises the issue of lesbian housing workers and the problems they face. So far lesbians have been looked at as 'housing consumers'. Lesbian housing workers face the same discrimination and may be reluctant to come out at work. Lesbian workers at the CHAR conference said:

> Experiences at work ranged from coming out with dreadful consequences or coming out and being accepted at work but not invited to socialise with co-workers, to staying in the closet because workers felt at risk from other workers or clients.
>
> (Cookson, 1989: 13)

Lesbians who do come out are often expected to deal with any lesbian and gay issues that arise and be the ones to challenge heterosexist and homophobic behaviour. Out lesbian workers can be a help to lesbian clients and the conference noted:

> It might be necessary to appoint lesbian and gay workers – at least initially – to get a response from lesbians and gay men who are homeless, being harassed etc. because lesbians and gay men are, quite rightly, apprehensive about coming out in such situations.
>
> (Cookson, 1988: 10)

A lack of succession rights can also lead to lesbian partners becoming homeless. Harrogate B.C. *vs.* Simpson (1984) ruled that a lesbian could not be considered a 'member of the family' even though she had lived with her lover for 20 years and so could not succeed to her lover's tenancy.

Bereavement can be especially hard for lesbians when other people do not acknowledge the significance of a relationship. Lesbians may face homelessness at this time too, through lack of succession and property rights which only recognise blood and marital ties.

The law is thus defining who has a legitimate right to live with whom. Carrie Harrison proposes that :

> Succession would not be limited to marital or blood ties, but to anyone who had lived in the accommodation with the tenant as their home for the previous six months. This would ensure equal rights for single people and lesbians and gay men.
>
> (Darke, 1992: 41)

This would go some way towards recognising self-defined households.

Whilst it could certainly be argued that lesbian couples should have the same rights as heterosexual couples, and that heterosexual couples' rights should not be dependent on marriage, this may not be enough. Housing policy dictates that people live alone or in households linked by blood or (hetero)sexual ties. Even living alone is not recognised as a valid life-choice. Single people are allocated bedsits because it is seen as a transient part of life. The lack of priority and good quality housing for single women constitutes indirect discrimination against lesbians who tend to remain single longer.

Many lesbians and gay men make a positive choice to live with friends but do not have access to social housing on this basis.

Talking of communal living among lesbians in the 1970s and 1980s Egerton quotes a lesbian who squatted at that time: 'We felt the struggle for lesbian housing was part of the same struggle as for single women's housing and for all women to have the right to choose how they would live and have autonomy' (Egerton, 1990: 82). Lesbians are still expressing a demand to share with other lesbians, as long as they can choose their co-habitees. The women's caucus at the CHAR Lesbian and Gay Housing Conference also suggested that sharing was one of the ways women obtain security to protect themselves against harassment. Another was: 'tending to gravitate to certain areas where they know other lesbians are living' (Cookson, 1989: 7).

There has been some debate in the literature about 'community vs. ghettoes'. Whilst some lesbians would welcome the opportunity to live in a lesbian community, others feel that ghettoes imply isolation and poor conditions. Lesbians may be prepared to put up with poor housing because of the opportunity to be with other lesbians:

> The fact that lesbians are forced to accept bad housing in the form of short-life property, squats, or 'hard-to-lets' on large council estates is indicative of society's intolerance of lesbians and gays. . . . Positive action to house lesbians in decent standard housing . . . indicates genuine acceptance not only of our life choices but also of our value as people. And such housing would help to create a safe and supportive community.
>
> (Anlin, 1989: 31)

However, one respondent summed up the debate: 'Could ghetto and community be the same thing but looked at from different angles' (Anlin, 1989: 30).

As a result of the oppression which lesbians experience in life, they may be reluctant to come out to housing workers, even if their housing problems are directly related to their sexuality. An immediate worry is, will the interviewer become hostile or refuse to help? A further concern is the confidentiality of the information given; how far will this information go?

Housing departments and individual workers can work to provide a safe environment for lesbians to come out in by displaying publicity for lesbians and gay men, challenging homophobic behaviour in reception areas and not making heterosexist assumptions about clients.

In addition to this, lesbians need to be reassured of the confidentiality of any information given. The AIDS and Housing Project (1993) has developed guidelines that talk of enabling informed consent to disclosure, so that people know where information is going, and can then decide what to say.

Fears about lack of confidentiality lead to lesbians' housing needs remaining invisible and perpetuate the not-an-issue-here syndrome.

EQUAL OPPORTUNITIES POLICIES TOWARDS LESBIANS AND GAY MEN: THE RESULTS OF A SURVEY OF SIX HOUSING ASSOCIATIONS IN JULY 1992

A key recommendation of the NFHA Equal Opportunities Report states: 'Associations should include lesbians and gay men as one of their official target groups for equal opportunities and have monitoring systems to monitor the implementation of policy in relation to lesbians and gay men' (NFHA, 1992: 8). To find out whether this recommendation had been implemented on a local level a questionnaire was sent to seven associations with properties in that area, and six replies were received.

All six had adopted or revised equal opportunities policies during 1990 and 1991. Whilst four of the associations included sexuality within the policy, only one had actually discussed the issue. One association stated that its policies did not extend to lesbians and gay men. Another had ambiguous wording in its policies, stating 'we will treat all people equally' but omitting lesbians and gay men from the list of groups of people experiencing discrimination. Lesbians and gay men might be unsure as to whether their existence was acknowledged under 'all people'.

Overall, four of the six associations surveyed were nominally targeting lesbians and gay men compared to 41 per cent in the NFHA survey (NFHA, 1992: 4). However, if the associations have adopted 'model' policies without discussing the issues with lesbian and gay groups and with their staff, it is doubtful whether they will be effective in implementing anti-discriminatory policies. An analysis of the results of this exercise is set out below.

Access to housing

All the associations have open waiting lists and accept nominations from other agencies, principally local authorities, but also various voluntary organisations. All would accept applications from lesbian and gay couples and from friends wanting to share accommodation. The stock profile suggests little availability of 'shared housing'. Housing policy generally does not recognise the validity of a choice to live with friends, seeing them as just so many single (low priority) people together. This is also reflected in housing design; for example, bedroom sizes based on a large bedroom for the parent(s) and substantially smaller other bedrooms which would not be appropriate for a number of adults living together.

Nomination rights rest increasingly with local authorities through partnership agreements. As authorities try to deal with the large numbers of statutory homeless households it is likely that lesbians and gay men (and other 'non-traditional' households) will remain marginalised in allocations.

Four associations, including the one whose policies did not extend to them, did allocate tenancies to lesbian and gay couples, and one said it had 'never arisen'. Only one had a date for the introduction of this policy (1990), and another commented that there was 'no specific policy'. Despite the specific exclusion from succession rights following Harrogate B.C. *vs.* Simpson three associations would grant new tenancies to a surviving partner. One association would not but 'would encourage a joint tenancy at the beginning', while another commented that 'it would be considered by committee'. The last association gave the familiar 'never arisen'. It is possible to argue that, in the absence of any publicity promoting such a policy, a surviving partner is unlikely to ask for a new tenancy to be granted.

Relationship breakdown

Housing policies in cases of heterosexual relationship breakdown vary enormously, so associations were asked simply whether lesbian and gay relationship breakdown would be treated in the same way. Four said that they did treat them in the same way, in one case it had 'never arisen' (again!), and one said 'no'. This last association rehouses a partner with childcare responsibility following a breakdown in a heterosexual relationship. Without any further details it is not clear whether there is a policy decision against lesbians and gay men, or just an assumption that they don't have children and therefore wouldn't qualify for consideration.

Confidentiality

As coming out, or being 'outed', can have serious consequences, confidentiality is an important issue for lesbians. This question was asked: 'Many lesbians and gay men do not disclose their sexuality beyond a circle of close friends as they are worried about the effect of disclosure on their working and/or personal lives. How do you deal with the issue of confidentiality?' One association answered that it had 'never arisen' but the other five simply answered that all information was treated confidentially. It was expected that information would be confidential within organisational boundaries, but it was important to know how confidentiality operates within the organisation. Is it necessary for information to be available to all departments, or even to all workers within one department?

An example could be where a lesbian, who had just been rehoused after experiencing harassment at a previous address, asks for extra security measures in her property. It is not necessary for the maintenance department or even outside contractors to know that the work is being done because the tenant is a lesbian; yet maintenance officers have been overheard discussing the sexuality of a lesbian tenant with a contractor. Letting a worker into her home is the contact with an association where a lesbian may feel most vulnerable. The answers given by lesbians later show that they have little confidence in confidentiality policies.

Training

Training is a key issue for the effective implementation of equal opportunities policies. The NFHA model equal opportunities policy includes the statement 'Besides the statutory duties not to discriminate, all staff have a responsibility to ensure the positive application of this policy' (NFHA, 1992: 27). It is necessary to receive some training to raise one's awareness of the issues facing those who experience discrimination, and to learn how to challenge such discrimination both in oneself, in others and in institutional practices.

Training in equal opportunities issues is provided by four of the associations through a variety of in-house and external courses. However, while four associations target lesbians and gay men in their housing policy, only two are sure that issues of sexuality and heterosexism are addressed in training courses.

Monitoring and targets

The NFHA survey recommends the use of monitoring systems to see how effective equal opportunities policies are in practice. It found that, while 41 per cent of associations targeted lesbians and gay men in their policies, only 2 per cent of associations monitored lettings, job applications and appointments by sexuality, and only 4 per cent of the waiting list and the sexuality of committee members (NFHA, 1992). Comparing monitoring with stated policy on housing management it noted: 'For women, people with disabilities and lesbians and gay men associations did not have any way of checking whether their policies were working' (NFHA, 1992: 7).

Information about monitoring by gender and sexuality was requested to get an internal comparison for each organisation. It was expected that at least allocations and job applications would be monitored by gender. Only two associations monitored rehousing applications, allocations, job applications, employees and membership of the management committee for gender, and one association monitored the gender of job applicants only. The other three associations did not monitor any of these for gender.

None of the associations monitored any area of operation for sexuality. Only one association answered the question: 'Can you see/have you experienced any difficulties with monitoring?' The answer was simply 'No'.

None of the associations had any target figures for lesbians and gay men in allocations, recruitment and membership of the management committee.

Monitoring cannot stand in isolation from a comprehensive strategy to implement equal opportunities policies. However, without any form of monitoring, the effectiveness of policies can only be measured by anecdotes and associations are not confronted with the actual outcome of their policies.

Although four of the six associations include sexual orientation in their equal opportunities policies, there are areas where the implications of this have not been worked through, particularly in training and monitoring the effectiveness of policies. The 'never arisen' answer is likely to continue unless the association takes an active role, e.g. contacting local lesbian and gay groups to discuss the issue.

It is not enough simply to add sexuality to the list in the policy. What is needed is a comprehensive strategy to implement policy, involving training and monitoring so that all workers understand the purpose of the policy.

THE HOUSING NEEDS OF LESBIANS: THE RESULTS OF A SURVEY CARRIED OUT IN 1992

The current housing situation of lesbians in the town was also of interest, as were any problems, costs, experience of the local authority or housing associations, and what sort of housing choices lesbians would like.

Britain is not a safe society for lesbians to come out in, and so it is not possible to carry out a survey which is representative of all lesbians. Lesbian existence is generally ignored or made invisible by heterosexual society, so there is no way of knowing how many women identify as lesbian; perhaps the next census could include a question about sexuality as a basis to work from! The survey concentrated on the local lesbian 'scene', the network of social events and support groups organised by and for lesbians, and places where lesbians can meet in safety as lesbians. This represents only a part of lesbian experience and in this case the scene was predominantly made up of younger, White, able-bodied women.

The participants were self-selecting and thus there was no control for race, age, disability or class. No Black women completed the questionnaire, the oldest respondent was 46 and only one woman stated that she was disabled. There was a varied class mixture. Overall, 18 women completed questionnaires. As there is so little research into lesbian experience generally, and housing issues for lesbians in particular, there is still a need to collect information and this survey ought to be seen in that context.

The respondents live in a variety of different tenures but are under-represented in owner occupation (28 per cent) and over-represented in private renting (39 per cent). This is a reflection on women's general economic position; however, another factor is the proportion of students (22 per cent) who are more likely to live in private rented accommodation. All the local authority tenants (16 per cent) have dependent children and it is their role as mothers which has given them access to council housing. Half the women spent 25 per cent or more of their income on housing costs.

The survey sought to identify what sorts of problem lesbians have experienced, whether they look to the local authority for help and whether they felt that their sexuality was a factor in the problems. Half the respondents have been homeless at some time and two-thirds have lived in unsatisfactory conditions. However, only one-third approached the local authority for help. A typical reason for not doing so is: 'No chance of getting anything as a single person.'

Indeed, none of the women without children received any help. One was told: 'Single women have to be over 45 to be on the waiting list for council housing; suggest go back to husband!'

One-third of the women experienced housing problems because of their sexuality, both from people they were living with and from neighbours. A woman living in student accommodation was 'outed' and had to move as a result of the harassment she then suffered. Another woman living in shared accommodation says: 'One heterosexual woman recently announced that she thought lesbianism was ultimately unnatural. We have since talked this through but it created problems for me.'

Nearly all the women felt that lesbians have housing problems just because they are lesbians. The replies indicate an expectation of discrimination and harassment. 'If they are out, lesbians get attacks from neighbours. Because people don't want to live with/ near us.'

It is interesting to note that, while 33 per cent have experienced problems, an overwhelming 89 per cent anticipate problems for other women. It might be thought that lesbians are overestimating the likelihood of discrimination; however, the possibility of verbal and/or physical abuse in a homophobic society is a real threat whenever lesbians are living as out and identifiable lesbians. The reason for this difference probably lies more in the behaviour of individual lesbians. In much the same way as women generally curtail their own behaviour to avoid potential danger, e.g. not walking home alone after dark, lesbians also have avoidance techniques. They do not walk hand in hand with their lover down their own street; they do not kiss another woman in their garden;

they take down lesbian posters when somebody comes to read the gas or electric meters. Such behaviour becomes second nature but it involves the loss of freedom to live openly as a lesbian.

The Lesbian Information Service study, *Lesbians and Housing in Leicester* (LIS, 1988), found that participants did not identify their housing problems as relating to their sexuality initially, but detailed case studies brought out the links.

Half the women have applied for public sector accommodation but only those with children have received help. Replies indicated that childless women feel that there is no point in applying. Although two-thirds of the associations surveyed do have an equal opportunities policy towards lesbians, 17 of the 18 women did not know whether there were equal opportunities policies, and the only women who knew that joint tenancies could be allowed actually had them.

There is clearly a lack of effective publicity on the part of the local authority and associations. If policies are to be more than paper commitment, this is something which should be dealt with in order to encourage lesbians to apply. A question about coming out to housing workers showed a great reluctance to do so. Here are some responses from lesbians who had applied:

Wouldn't do it.
Although I am an out lesbian, I do not think I need to state my sexuality to anyone unless (a) I want to and (b) it is relevant.

From lesbians who had not previously applied for housing:

Wouldn't.
I would be scared it would jeopardise my application.
I would find it a very stressful event.

Coming out to anybody in 'officialdom' is very different from coming out to friends, relatives or even work colleagues. Officials keep records, which may be open to other people. Also, officials have more power than you do, regardless of how they deal with that power imbalance. The answers above show how important it is for workers to receive training in awareness of sexuality issues and heterosexism in order to provide a safe environment for lesbians to come out in.

Only 28 per cent of respondents would have no worries about the confidentiality of information given to housing workers, and 50 per cent would definitely worry about confidentiality.

It is no surprise that all 18 lesbians think lesbians and gay men should have the same succession rights as heterosexuals, with a slightly lower number saying that this should also apply to friends who have lived together.

The research sought to explore how lesbians would like to live, whether being part of a lesbian community is important and whether sheltered accommodation is seen as a viable option in old age. Living near other lesbians is an important consideration for two-thirds of the women.

Many lesbians share with friends or live in other people's houses owing to financial constraints. Ten women would still choose to share accommodation, as long as they could choose their co-habitees.

None of the women wants to live with men and most favour smaller households. When sharing is the only option because of financial constraints and there is no choice about who to share with, where accommodation is in short supply, lesbians are more likely to have to share with men. Associations operating sharing schemes should ensure that they offer women-only houses and consider lesbian houses too, as lesbians do suffer harassment from heterosexual women in shared accommodation.

The women were asked to describe their ideal and long-term housing situations. Most people do not design their own homes and, when thinking about where they would like to live, it is often things such as location, garden size and proximity to friends that influence them. In the same way these answers reflect issues that are important to lesbians but do not describe the 'bricks and mortar' at all. The 'ideal home' fell into three main categories:

No discrimination
One where women don't get stared at if they walk down the street hand in hand . . . I don't want to live one life in my house and another one outside – maybe I'll give up this attitude in a few years' time!

Lesbian community
A completely lesbian community.
In self-contained accommodation with other lesbians around.

Security
In my own home.

The long-term choices also fell into broad categories:

Moving to the countryside
Yes, live in the country but with a largish lesbian network nearby.

Living communally
I want to buy a house with a lesbian friend or share housing association house with a lesbian friend, in an area with other lesbians.

Owning property
Yes, owning my own home.

Two-thirds of the women say they would consider moving into sheltered accommodation. The factors influencing the 'yes' choice emphasised the importance of being accepted as a lesbian:

Whether my partner and I were accepted.
Only if there were other lesbians there and the warden was a supporter.

One respondent interpreted the question to be about an 'old dykes' home':

Knowing I would not be so isolated living around lesbians as we are all getting older, with support.

For the lesbians who would not consider this move, loss of independence and lack of acknowledgement of lesbianism were the key factors.

Although in practice many sheltered schemes and residential care homes have a majority of women residents, this is not the same as a positive choice to live in a women-only setting. Sixteen lesbians felt that there should be women-only provision. This would respect the choices which women have already made during their lives, rather than asking women to accommodate themselves to the way in which care is provided.

Overall, the key elements of housing choice are to be able to live in an area with lesbians (66.5 per cent); to live with women (100 per cent of those wanting to share); and to live without fear of discrimination.

CONCLUSION

> Expectations of good practice require associations to target
> these policies not only to black and racially discriminated groups
> but equally to women, people with disabilities and lesbians and
> gay men
>
> (NFHA, 1992: 1)

A review of the National Federation of Housing Associations'
report (NFHA, 1992) indicates that lesbians and gay men are
always the least-targeted target group. The report comments that
'relatively few [associations] are using any monitoring in terms of
women, disability or lesbians and gay men' (NFHA, 1992: 7). This
should be seen against the fact that 41 per cent of associations in
their survey targeted lesbians and gay men in their overall equal
opportunities policies (NFHA, 1992: 4), and '46 per cent of the
same . . . were "neither satisfied nor dissatisfied" with their progress on equal opportunities' (NFHA, 1992: 17), and '48 per cent
of small associations said they were very satisfied or satisfied with
their progress on equal opportunities' (NFHA, 1992: 17).

The report goes on to give nine examples of current practices.
Four of these are specific policies in respect of race and disability
issues and one is a model equal opportunities policy. Three others
are more general in their aims and specify combinations of other
groups but do *not* mention lesbians and gay men. The only good
practice example that mentions lesbians and gay men at all is one
specifically about them.

It is important to realise that any proposals for action are made
in a context of constant funding cuts and attempts to remove
housing from the public sector altogether. However, it is surely
not the case that 'crisis management' removes the responsibility to
work towards equal opportunities, and not all recommendations
have funding implications but are about ensuring that lesbians
(and gay men) have the same access to resources as heterosexuals.

The most obvious recommendation is, of course, that housing
providers adopt policies which specifically target lesbians.
However, paper policies alone are not enough, and policies should
only be adopted after discussion of the issues. To be implemented
effectively such policies must be backed by training for all staff,
management committee members and tenants' groups.

The practical implications of the policies also need to be worked
out, e.g. points systems that recognise the harassment experienced

by lesbians, allocations policies that accept self-defined households, and how to publicise policies effectively to lesbians. Maintenance departments should also take positive action to ensure that they employ and/or train women maintenance workers so that lesbians have a choice and can ask for a woman to do a repair.

The Housing Corporation does not go as far as to utter the words 'lesbians and gay men'. Its performance expectations: 'require associations to . . . ensure equal opportunities in all activities for . . . others vulnerable to discrimination' (NFHA, 1992: 1). It is entirely consistent with state policy not to recognise lesbians' existence.

Implementing effective equal opportunities for lesbians is not simply a matter of noticing a previously unmet need and getting on with it. If a government can pass Clause 28 they are not going to introduce 'sexuality equality' legislation. Equal opportunity housing policies for lesbians go against state and religious ideology and against the view of many people. Clause 28 was reputedly introduced because lesbians and gay men were becoming too visible, 'out and proud', and it would not be at all surprising to see a backlash against housing providers trying to implement equal opportunities for them, e.g. being 'less favoured' in their funding bids, or guidelines that 'lesbian and gay rights have no place in housing policy'.

To conclude, these words from Jayne Egerton summarise the situation: 'I believe that the most disturbing aspect of the last ten years has been the extent to which the material conditions which make a lesbian life and identity possible are being progressively undermined . . . The struggle has never been simply about bricks and mortar' (Egerton, 1990: 86).

NOTE

1 The title for this article is taken from Jayne Egerton's article in *Feminist Review*.

REFERENCES

AIDS and Housing Project (1993) *Confidentiality Policies and Guidelines*, London: AIDS and Housing Project.
Aimsworth, P. (undated) *A Report into the Needs of Older Lesbians and Housing*. Unpublished, for Pensioners Link.

Anlin, S. (1989) *Out But Not Down: The Housing Needs of Lesbians*, London: Homeless Action.

Cookson, S. (ed.) (1989) *Housing for Lesbians and Gay Men*. Report of a conference held on 1 October 1988, London: CHAR

Darke, J. (ed.) (1992) *The Roof Over Your Head: A Housing Programme for Labour*, Nottingham: Spokesman.

Dibblin, J. (1988) 'Jenny lives with Eric and Martin', *Roof*, November/December: 25–7.

Dibblin, J. (1991) *Wherever I Lay My Hat – Young Women and Homelessness*, London: Shelter.

Egerton, J. (1990) 'Out but not down: lesbians' experience of housing', *Feminist Review* 36, Autumn. (N.B. Title acknowledged as that of Anlin.)

Foster, J. (1988) 'The dark side of the boom', *Youth in Society*, January, 20–1.

Harrogate B.C. *vs.* Simpson (H.L.R. 17) 1984.

Labour Research Department (1992) *Out at Work: Lesbian and Gay Workers' Rights*, London: LRD Publications Ltd.

Lesbian Information Service (1988) *Lesbians and Housing in Leicester. A report by the Lesbian Information Service*, Leicester: Lesbian Information Service.

National Federation of Housing Associations (1992) *Equal Opportunities in Housing Associations. Are You Doing Enough?*, London: NFHA.

Watson, S. and Austerberry, H. (1986) *Housing and Homelessness: A Feminist Perspective*, London: Routledge & Kegan Paul.

FURTHER READING

Barratt, M. (1988) 'Advising lesbians and gay men', *Roof*, November/December: 38–40, Housing Advice Notes No. 78.

Witherspoon, S. (1991) 'Housing issues for lesbians', *Adviser*, Issue 27, September/October: 23.

Chapter 9

Snakes or ladders?
Women and equal opportunities in education and training for housing
Marion Brion

INTRODUCTION

What can education and training do to improve equality of opportunity in housing? What should the aims of such intervention be? There are a number of possible aims which can be summarised in three main areas:

- to establish an understanding of how prejudice and discrimination function to influence life-chances and housing outcomes;
- to demonstrate effective ways in which the housing service can respond to equality issues;
- to enable women working in the housing service as employees, committee members or volunteers to develop their potential in relation to that work and improve access to such opportunities.

Many of the issues related to women's access to housing, suitability of provision and women managers are discussed at length in other chapters. Part one of this chapter will therefore concentrate on an overview of the main constituent characteristics of disadvantage to women in housing work. The other issues related to gender and housing will be summarised.

It is impossible to ignore the fundamental significance of general education at school level and above. Part two will identify the main areas of disadvantage in general education and in relation to housing education and training, and discuss some of the remedial measures taken. Part three draws on the experience of the past 12 years in housing and education to present a checklist which employers and housing staff can use in order to achieve the aims outlined earlier. It illustrates the application of the checklist at each level. Readers who want this practical level can therefore pass directly to Part three but I would argue that we can all make a better case for equal opportunities when we have evidence to back our case and understand some of the theory.

PART ONE: OVERVIEW

The issues in housing

Brion and Tinker (1980) delineated some of the disadvantages of women in housing employment using an analysis of Institute of Housing statistics and data from the City University staff study (1977). Subsequently, two large studies, *Women in Housing Employment* (NFHA, 1985) and *The Key to Equality* (Levison and Atkins, 1986), established beyond doubt the basic characteristics of women's disadvantage in housing employment. Figures 9.1 and 9.2 illustrate the main points. Women were employed in large numbers by the housing service but were concentrated on the lower grades. They were also disproportionately over-represented in some functions and under-represented in others (see Figure 9.2). These figures illustrate the pattern of horizontal occupational segregation (grade and salary levels) and vertical occupational segregation (types of work) identified by researchers such as Hakim (1979) who have studied general characteristics of women's employment. The evidence was sufficient for both the National Federation of Housing Associations (NFHA) and the Institute of Housing (IOH)[1] to take action and their reports contained official recommendation for action by employers, and by the NFHA and the IOH.

Getting implementation of effective equal opportunities action, both inside and outside these organisations, took much longer. The process is more fully described in *Women in the Housing Service* (Brion, forthcoming) but some aspects are included in Parts two and three. During the 1980s many other studies of women in relation to housing were produced and the analysis developed to include issues such as race, age, class and disability (e.g. Rao, 1990; Morris, 1988).

One key question, therefore, must be: 'How far has the position of women in the housing service improved after nearly a decade of action?' The honest answer is, regrettably, that we do not know but we can make some educated guesses.

The most recent equality survey is one carried out by the NFHA

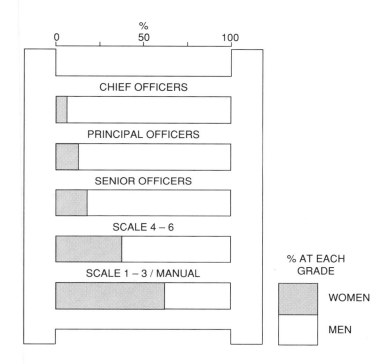

%

0   50   100

CHIEF OFFICERS

PRINCIPAL OFFICERS

SENIOR OFFICERS

SCALE 4 – 6

SCALE 1 – 3 / MANUAL

% AT EACH GRADE

WOMEN

MEN

Figure 9.1   The career ladder 1986.
Source: Levinson and Atkins (1986)

in 1990. However, this was a general survey on equality and was not designed to be comparable with the 1984 NFHA women's equality survey. The differences between the two are sufficient to make any comparison unreliable. Unfortunately, an illusion was generated that things were substantially better for women.

The view that some things were getting better, to some extent, was more reliably indicated by the IOH statistics. Table 9.1 shows these in relation to long-term trends. It is clear that the aim of getting appropriate proportions of women to enter the profession and qualify is now being met. For some years representation at Member and Fellow level has remained out of line, indicating women's continued problems in proceeding to senior posts. However, the 1993 statistics show a greater proportional increase in female Fellows, so it is possible that the

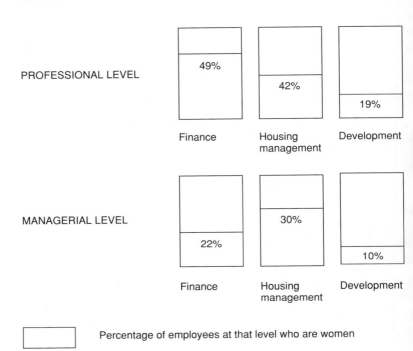

Figure 9.2 Comparison of the proportion of women at professional and managerial levels in three functions of housing.

Source: Survey of Housing Associations 1984 (NFHA, 1985)

increase in recruitment is finally feeding through to the higher grades. It should be noted, however, that the proportion of women Fellows is not yet even back to its 1965 level. From the end of January 1992 to the end of January 1993, female Fellows increased by 21 per cent, while male Fellows increased by 12 per cent. IOH information only applies to those who are, and continue to be, members of the professional organisation and does not cast any light on vertical occupational segregation.

There is a danger that a premature complacency may set in. Where the figures do represent improvement, they may be the results of equal opportunities work in the early 1980s and may not be sustained in the harsher conditions of the 1990s. Without employment surveys, which are expensive, it is difficult to cast any

Table 9.1   Institute of Housing membership

	1965		1977		1983		1993		Total
	M	F	M	F	M	F	M	F	
Fellows	79%	21%	85%	15%	89%	11%	81%	19%	
	(223)	(61)	(224)	89	270	32	816	190	1006
Members	79%	21%	82%	17%	77%	23%	57%	43%	
	713	189	952	201	1250	367	2479	1852	4331
Students	85%	15%	76%	24%	60%	40%	45%	55%	
	837	142	1728	535	1398	1076	1940	2376	4316

Source: Chartered Institute of Housing

light on this. Fortunately, it was possible to extract data from NFHA training records which produced some additional evidence. Concentrating on the two areas where disadvantage to women has already been demonstrated, Figures 9.3 and 9.4 suggest that in 1992 both vertical and horizontal occupational segregation in housing are only too alive and well and illustrate some of the mechanisms of disadvantage. They confirm anecdotal evidence (emerging, for example, at the NFHA Women's Conference sessions and at fringe sessions at other conferences), that women are still suffering from considerable discrimination, particularly in relation to higher grade posts and technical functions of housing. They also tended to substantiate evidence from individual women that certain conferences were not 'woman friendly' and led the NFHA to investigate this issue further (NFHA Women's Standing Group, 1992).

The greater emphasis given to equality of opportunity in the Housing Corporations' performance criteria in 1992 was another welcome sign. But central government support can vary with the political wind. The DOE efficiency report, *Training, Education and Performance in Housing Management* (DOE, 1990), did not mention equal opportunities and referred to the housing officer as 'him'. Government support for Opportunity 2000 is helpful but there are fears that with Compulsory Competitive Tendering narrow efficiency criteria may again predominate.

The indication from this data, that there is now a long-term progression in which larger cohorts of women are moving through

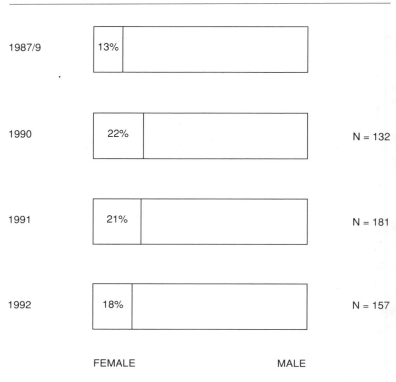

Figure 9.3   Attendance at NFHA chief executives conferences 1989–92.

the different grades, is therefore encouraging. But this encouragement must be tempered with an awareness that many of the
barriers to women's access to power are still in place. The achievements of the 1980s should form a basis to build from rather than a
summit.

Women as committee members and volunteers; women and tenant participation

There are many women working in the housing service who are
not paid for what they do. However, the relative participation of
men and women in these roles has not been studied as much as in
the employed sector. The 1984 NFHA survey stated:

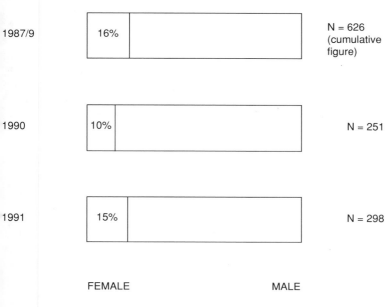

Figure 9.4   Attendance at NFHA maintenance conferences 1989–91.

It is startling to find that women seem also to have been dis-
placed on committees of housing associations, especially on
those with a large stock. 43 per cent of associations had over 80
per cent male membership on the committee . . . a total of 67
per cent of associations with committees where male membership was more than 70 per cent.

(NFHA, 1985)

Research for the Jubilee Working Party indicated that the position
tended to be rather better in smaller associations. Recommendations were made for improving the balance as part of an overall
aim of improving the representative nature of committees. Work
in the past on the participation of women in local government has
tended to indicate that the same factors which disadvantage them
in employment also disadvantage them in voluntary work.

Kearns' research in 1990 again found an under-representation of
women on housing association committees of management (30 per
cent female, 70 per cent male, as against 51.9 per cent women in
the relevant population). Kearns judged that this represented little

change on the situation described by Crook (1985). It seems likely that the distribution of posts of chairs of committees is also heavily gender biased but there is little direct information on this point (Kearns, 1991).

Given the current stress on tenant participation and tenant management, the participation of women in tenants' organisations is important. The issues outlined below apply to women committee members and tenants as much as they do to employees.

Women's access to housing and the suitability of provision

A number of chapters in this book illustrate how gender affects access to and provision of housing and how the needs of many groups of women have been overlooked in the past. Although the issues are complex it is helpful, from the point of view of this chapter, that the underlying causal factors are similar to those which we will be considering in relation to educational processes.

PART TWO: EDUCATION AND TRAINING

A major task of the women's movement has been to develop awareness of the way in which ideas about appropriate roles for men and women are developed. While there is still some controversy about inherited characteristics (Garrett, 1987), there is massive evidence that ideas about what men and women can or should do are formed initially in early childhood socialisation and by the media (see, for example, Connell, 1987). The educational system can then heavily reinforce stereotyped ideas. These stereotypes might include the idea that men and women have innate psychological differences and aptitudes for different kinds of study or occupation, the assumption that men are to be regarded as the norm from which women deviate, that men 'naturally' assume positions of leadership and power, that women's careers will be distorted by the overriding desire to have a family, and that women will be the carers not only of children but also of others such as the disabled and the elderly (see, for example, Wolpe, 1977).

When demands for change began to be voiced, some of those involved in education argued that education always reflects society and that they should not be held responsible. But in general, subsequent to the Sex Discrimination Act, overt resistance was muted, while covert resistance continued and has reappeared with greater force in the 1990s.

While there is considerable debate about the origins of discrimination on the grounds of gender, there is agreement that in the past the educational system has reinforced such ideas. It is useful to consider three levels of influence:

- *Institutional level* – the unequal distribution of men and women in relation to positions of power within education and training; inequalities in access and provision.
- *Behaviour level* – differential treatment given to males and females in the educational process.
- *Content level* – general influences in the content and written language of teaching and educational materials which reinforce stereotyping.

While there has been, and will continue to be, argument about the proportional contribution of each of these to women's disadvantage there is no doubt that together they constitute a system which works to create and sustain disadvantage. They have a cumulative and important effect.

> Discrimination by race and sex in the job market is largely a result of the sex and race stereotyping and discrimination in education and training programs, which restrict women's occupational options and are based on outmoded notions of women's proper roles in the home and workplace.
>
> (Wolfe, 1991)

Even beginning to identify all the different mechanisms involved shows also that they are complex and entangled together. Wolfe comments:

> this topic is in an incohate stage . . . the policy worlds of education and of employment are difficult to grasp as individual systems, let alone as connected policy labyrinths. And when issues related to gender, class, race and ethnicity are added to this mixture the result for some is deadlock.
>
> (Wolfe, 1991: 201)

This chapter is initially concerned with gender issues and there are stages when one source of disadvantage needs to be separated out so that it can be studied properly. But if we believe that we cannot build equality for one group on the basis of inequality for others, other inequalities need to be considered. There are stages in the implementation of equality programmes when it is essential to have policies which are comprehensive and coherent. An awareness of the complexity of the issues need not produce deadlock if it leads us to consider that this is a system and, if we seek to change one part of the system, it is advisable to see what corresponding changes may be needed elsewhere.

In this chapter some of the complex issues must necessarily be dealt with by simplification, summary and an indication of where they are more fully discussed. Concentrating on the relationship to the housing service also helps in dealing with the complexity. Other writers (for example, Connell (1987)) focus clearly on the issue of power and this approach will also be used as a thread through the labyrinth. When it is remembered that the purpose of the system which we are studying is to maintain privileged men in power, its complexities and contradictions become easier to understand.

The next section will now explore each of the areas in a little more detail.

The institutional level: power and stereotyping in the staffing of the educational system, in access and provision

Early studies demonstrated that the distribution of staff within the educational system reinforced stereotyping and that positions of power were generally held by men. For example, the domination of head teacher or principal posts and other senior jobs by men reinforced the idea that leadership and management belonged to men. Thus men had the power to defend their position, while younger women lacked role models who would help them to envisage such a career. The proportion of women in senior posts in education actually fell in the 1970s as educational organisations grew larger and the jobs became better paid (Byrne, 1978).

The domination of technical, maths and physics teaching by men reinforced the tendencies of girls at primary and secondary levels not to choose these subjects, and this even began to be felt in the new field of computing despite women's ability to grasp the technical complexities of word processing (McGivney, 1992).

In further and higher education both these areas of imbalance become worse. Equal opportunities legislation has tended to have its greatest effect on recruitment and promotion procedures. Despite this and despite efforts by unions and some employers, progress has only been limited (McGivney, 1992).

It is interesting to take an example from a subject which is not generally regarded as technical and is studied equally by men and women. In 1991 a study of university teaching posts in history, a subject which nearly equal proportions of men and women read as a first degree, revealed that only 17 per cent of lecturing jobs in a sample of 53 universities were held by women. The proportion declined to 12.7 per cent among senior lecturers, and only three out of 134 professors. One conclusion was that history had a more 'male' image than appears at first sight. Another was that the 1990 increase in the proportion of female appointments might be due more to academic appointments becoming less attractive to men through uncompetitive salaries than improved equal opportunities practice (Griffiths, 1991). This unfortunately illustrates all too well how long-lasting and pervasive discrimination in institutions and society is.

Checking the statistics of staffing in relation to grade levels and types of work remains a vital aspect of equal opportunities monitoring. Direction from the top is of crucial importance. It is essential, therefore, that college governors or managing boards and local TECs also reflect equal opportunities staffing. Now that educational institutions are becoming independent this is one of the items which must be monitored.

The collection of statistics on students by gender, ethnic origin and type of course is also extremely important. It will indicate how far the institution is being successful in breaking down barriers of access to certain types of occupation or higher paid jobs for under-achieving groups. Information thus gathered can lead to action in relation to improving mainstream provision and running special courses, such as access or fresh-start courses for women. The collection of data on enquiries and applications as well as student entry will indicate whether the problems are in publicity or in the applications process.

Housing and related courses

At present the IOH cannot offer a gender breakdown of staffing of housing courses. Anecdotal evidence suggests that there has been a mixed evolution on the core staffing of housing courses. In the early stages, when courses were often located in building or construction departments of further education (FE) colleges or polytechnics, there was a predominance of men. As courses grew, more women were involved as course leaders as well as lecturers, particularly in low status FE colleges. When courses shifted to polytechnics and universities, the proportion of men increased, particularly at course leader level. There is some historical evidence which would enable partial checks to be made if the IOH and the educational organisations decided to monitor this data.

If an apparently neutral subject like history can turn out to have a male image, it is all the more easy to appreciate the disadvantages which women have encountered with regard to the design, construction and repair of buildings. In our culture, building has been seen as definitely a 'male' occupation (despite the fact that in other cultures it is women who are the house builders). Research has shown that the closer an occupation is to the building industry, the lower the proportion of women (Brion, forthcoming). This affects women who want to work in those functions of housing or in related committee work, or as surveyors, planners or architects. It can also affect access to jobs at the top of housing organisations, as design or technical experience may be seen as a valuable asset in the promotion stakes. Finally, this issue also affects women who are home owners or tenants responsible for their own repairs. This is clearly a major issue which neither school nor post-school education is yet tackling satisfactorily, though there have been some encouraging initiatives.

Housing courses normally draw on specialists from such subjects as law, accountancy and building, so the influence of the image of related professions and building is relevant. Research has shown that in architecture 1982/3 teaching staff were 97 per cent male. In surveying, in a 1991 study, 'the percentage of women lecturers varies considerably from college to college from about 1 per cent to 10 per cent, with a few exceptional colleges with around 20 per cent' (Greed, 1991). Greed found that some exceptional women had been able to make their way in surveying education, partly because in times of full employment academic salaries were less than what could be gained in private practice, and status in surveying came from successful practice rather than academic work. However, this limited number of exceptional women could still not provide students with the choice of role models they needed (Greed, 1991: 97), and the prevailing culture of the department could still be very male dominated. Other women could be employed in part-time posts, leaving control of the department with the men in full-time employment (Greed, 1991: 98). The effect on women students of such imbalances in staffing can be profound. They often lack an appropriate choice of role models as well as gender-appropriate counselling and support. It increases their difficulties in dealing with discriminatory behaviour.

One response to these problems has been to set up all-women courses at access level or 'women into building trades' courses but this was not usually done for housing or surveying. However, some housing departments pursued enlightened policies of special support for female staff who wanted to progress in technical jobs and courses to upgrade the technical knowledge of staff, e.g. repair clerks, who were usually female and working class. The Positive Action Training in Housing (PATH) schemes, which were intended to improve the proportion of Black and ethnic minority staff in housing, because there was an equal opportunities policy, also benefited Black women.

The IOH played a major part in the 1980s, both in creating a national framework for housing qualifications which maintained access for disadvantaged groups, and in providing return to work courses for experienced staff who had dropped out of housing because of a career break (see Part three and Brion, forthcoming).

It is clear that women in housing have some common concerns with women in surveying, planning, architecture and the manual trades. Women have made some efforts to draw these issues together. The persistence of inequality with regard to types of work indicates the need for ongoing and coordinated action.

Behaviour of staff and students

There is plenty of evidence that the behaviour of teachers and trainers towards male and female pupils or students can reinforce stereotypes and fail to give women equal opportunity to develop their potential. First, trainers at all levels tend to give more attention to males than females in mixed classes, and students themselves will often regard this as natural and will not comment on it unless the order is reversed. Second, research done by Lakoff (1975) and Spender (1980) has illustrated the way in which men dominate mixed sex discussions. This can be a difficult factor to deal with in any group which has set up its social norms. Third, staff and students can reinforce sexist attitudes by their behaviour (Spender, 1982). Fourth, there can be, and has been, sexual harassment by staff or students. Fifth, women's needs can be ignored by the way in which the curriculum is planned (Whylde, 1983).

In relation to housing courses there is evidence of all these types of behaviour, particularly (but not only) in relation to the teaching of technical subjects. Instances of discrimination in the applications process have been reported to the IOH women's working party. Sexual harassment by lecturers or by other students has also been reported. Instances of sexist language and assumptions that women cannot cope with technical subjects have been numerous. One senior woman housing officer, who wanted to obtain a Higher TEC Certificate in Building, commented on the shock of finding herself 'isolated' in a class of men and the conflict engendered by letting sexist comments by staff and students go by. 'I just did not have the energy to stand up to them – it would be more difficult to survive if I had and it was a matter of survival'.

Areas of omission, like giving more attention to men than to women or allowing men to dominate discussions, are even more frequent. A senior qualified woman, chair of the housing management committee of a large housing association, commented that she had never yet attended a management course where the trainers seemed aware of this issue, and this included courses run by the NFHA.

In many institutions there was a tough battle to get greater acknowledgement and understanding of gender issues written into the curriculum and improved practice accepted by all teaching staff. Women in an institution with a stated equal opportunities policy might still find themselves made uncomfortable or ridiculed by other staff when they raised such issues.

Both the NFHA and the IOH have produced guidance for speakers and trainers following pressure from disadvantaged groups and the evidence from general research. This is one of the issues on which there has been a backlash – centred around the issue of 'politically correct' language. It is worth noting that this label itself is discriminatory. 'Language which allows disadvantaged groups to develop their potential' is too cumbersome but would be more helpful. Even so it must be admitted that some proponents of equal opportunities in language and behaviour have provoked adverse reactions. The parallel case of anti-racism training has demonstrated that the most effective strategies rely on changing the culture of the organisation by policies, procedures and guidelines, recruitment and promotion, close monitoring, targets and action plans, as well as training for existing staff. Continued monitoring of outcomes, e.g. proportions of students and drop-out rates and requiring those not meeting targets to put forward their own plans for improvement, takes the burden off individuals (Luthra and Oakley, 1991).

Because the NFHA and the IOH have had policies and procedures, most of the incidents described above have been dealt with. But women involved in raising issues of this kind are put under a great deal of stress and need support, which they do not always receive. Housing employers who are seeking to promote equality of opportunity and staff therefore have a key part to play. They can monitor behaviour which is much easier to do at the local level. Behaviour is a more sensitive issue to research than statistics of staffing but it is as important, and there are perfectly viable observation methods which can be used in small studies.

Employers or colleges can also provide various kinds of support and help for women returning to learning or challenging accepted stereotypes. In the 1980s the emphasis here was on assertiveness courses. These were useful but sometimes left women rather stranded with regard to ongoing development. Discussions at the NFHA women's conference have indicated that more broadly based support for those struggling for equality is needed, including suitable personal development and such courses as negotiating skills which are often seen as the prerogative of management or unions.

The general content of education and written language

Early studies showed that sexism in society was reflected in the content and language of the curriculum and educational materials and there has been a long, hard struggle to redress this. Replacing the sexist language and behaviour of 'Janet and John' and similar readers became a symbol of this in primary education. At higher levels and in relation to housing there is still some way to go. There is plenty of guidance available now on written language and it is an easy issue to monitor.

For many years there was a lack of material which analysed equal opportunity issues in relation to housing or which reflected the outlook and needs of women, ethnic minorities and disabled people. There is now much more available – illustrated by many of the other chapters in this book. The problem remains one of selecting the material and competing for a place in the curriculum.

Early approaches tended to locate equal opportunities in a few specified teaching sessions. Modern approaches emphasise integration with all relevant aspects of the curriculum – but this is more difficult to monitor. Many staff argue that it remains desirable to locate clearly an introduction to equal opportunities issues in order to produce an understanding of the structural inequalities. In recent years there has been a tendency for sociology or social studies to be pushed out of the curriculum in favour of economics and this has led to concern that teaching about inequality lacks its proper base and that understanding is fragmented. (Possibly this is as much a comment on economics as commonly taught as it is on sociology. See, for example, Waring (1989).)

However, there are now many good examples of ways in which information about the structures of inequality and improved practices of housing authorities can be incorporated in the content of teaching. But the behavioural processes described above are still often ignored. It was surprising in 1993 to find communications and management courses dealing with human interaction but 'gender blind', yet this is still happening. Employers and staff, once aware of the possibilities, can play an important part in monitoring this.

PART THREE: CHECKLIST

In the 1990s education and training providers will be working within a market economy. New systems such as NVQs should give employers more choice as to who delivers training. Employers who seek to promote equality of opportunity are therefore in a more powerful position to influence providers. Working from the experience of the past decade, I have produced a general checklist which can be adapted for use at any level and this has been piloted with various groups. It is divided into sections for use at the different levels: the institution, behaviour and the content of education and training.

The checklist can be adapted for use with internal and external training and awarding bodies as well as colleges. The first section will therefore demonstrate how the checklist can be applied to the award-giving bodies, at institutional level, before going on to look at local action.

Institutional level

In the case of housing education the main bodies involved are the National Council for Vocational Qualifications (NCVQ), the Business and Technology Education Council (BTEC) and the IOH. These organisations were first asked the questions on the checklist in 1991 and it has been possible in most cases to compare 1993 responses.

Table 9.2   Questions at institutional level

---

1 Does the institution have an equal opportunities policy?

2 Has the policy been translated into practical action?
  In particular:

3 Is responsibility for implementation and monitoring clearly located?
  Is there monitoring of staff and consultants by area of work and
  grade: gender, race, age and disability?
  Is there monitoring of students/trainees by type of course and level of
  studies: gender, race, age and disability?
  What action has been taken in response to monitoring?
  Is the onus put on the institution and its departments to produce
  plans for improving equality of opportunity?

4 Is there guidance on non-discriminatory practice for staff and
  students and training for staff in how to implement this?

5 Do arrangements for provision of courses, recruitment, admissions
  and publicity reflect an awareness of equal opportunities issues and
  effort to improve representation of target groups?

NCVQ

In both 1991 and 1993, NCVQ had an officer with equal opportunities as part of his duties. The NCVQ required awarding bodies to monitor equal opportunities but by 1993 it had realised that this requirement was not specific enough. National data were not gathered, though there was monitoring by sex and ethnic origin for some courses. Following a review of this issue, by 1993 a process of working with awarding bodies to produce common accord on the collection of statistics and on action in response to them had begun.

During the process of formulating the NVQ standards for housing there was an NCVQ equal opportunities working party, with representatives from the NFHA and the IOH, and careful attention was given to equal opportunities issues in this group. Nevertheless, when the draft standards appeared it was felt that they were not sensitive enough to equal opportunities. Further representations were made and revisions were then undertaken. Considerable effort had to be expended because NVQ standards are detailed and meetings were time-consuming but, since NVQs were intended to be a major route of access for those with less educational qualifications, it was felt to be important. This is an example of the way in which it is essential for those working for equality to be on the spot when changes are being made and pay attention to the detail of major issues.

BTEC

In 1991 BTEC did not have either an equal opportunities policy or a designated officer. The London branch of the IOH became concerned about BTEC's presentation of publicity – for example, a 1991 leaflet on design courses had a picture of one white male on the front. However, by this stage BTEC had already begun a review of equal opportunities. In May 1992 it had produced a policy. Subsequent to this date publicity has done more to promote quality of opportunity.

In 1993 BTEC monitored both students' enrolment and completions and its own staffing by gender and age, but not by race or disability. Equivalent statistics were not collected for moderators (external verifiers), who work on a contract basis though appointed by BTEC. Consideration was being given to extending monitoring, particularly in relation to the Common Accord on NVQs and with regard to staffing. For BTEC NVQs, moderators had a specific criterion to check in relation to equality of opportunity with a number of indicators specified. For existing BTEC courses they were required to ensure that BTEC's policies were followed, the only specific mention of equal opportunities checked being in relation to entry. A 1993 inquiry into completion rates was able to use the existing monitoring statistics to demonstrate that, in almost all awards, female students achieved a greater percentage of full awards than male students. This applied to areas where women traditionally form a small minority as well as to those where they are in a majority (Smith and Bailey, 1993).

The Institute of Housing

The IOH had had a women's working party since the early 1980s, which carried out initial work to improve the retention of women members and ensure that a basic equal opportunities policy and monitoring statistics were implemented. By the end of the decade this working party increasingly felt that there was a need to integrate equal opportunities action with the main policies and procedures of the Institute. In addition, concern about IOH responsibility for promoting equal opportunities in education came to a head with the joint DES/IOH conference in 1989, when five out of six speakers planned and all the workshop leaders on the first day were male. Later, in 1989, following discussions, the women's working party was dissolved and an equal opportunities Working Group with a clear brief as part of the new structure of the Institute was introduced. Among other activities a review of education and training was carried out, which demonstrated that the IOH had gone a considerable way in implementation and identified further action needed. In November 1992, the IOH Council accepted most of the recommendations of the working group. These included, for example, the target that 30 per cent of the members of Council should be women by 1995 and 50 per cent by the year 2000. It was agreed that the attainment of the target should be monitored and reviewed in 1994 with a view to further proposals being considered, if appropriate, in 1995.

A major achievement of the IOH, in cooperation with other organisations, educationists and housing practitioners, has been the formal 'ladder of opportunity'. This structure incorporates the BTEC National and Higher National Certificates as a route to professional education for non-graduates. There was concern, when the IOH was initially upgrading their qualifications in the early 1980s, that it would move towards an all-graduate entry and cut off part-time routes, like a number of other professions. Fortunately, the lobby against this did not rely entirely on those committed to equal opportunities. Many housing employers, of all political complexions, valued the contribution of staff who had 'come up the hard way'. The IOH also had an interest in maintaining a strong role in education.

Many of the existing teaching staff were committed to equal opportunities. Nevertheless, it took a great deal of time and cooperation from those concerned to achieve the new structure within the framework of interest groups involved in changing a national curriculum.

Both women and ethnic minorities were interested in maintaining the part-time routes. Although the college statistics are not complete there is already confirmation that these courses are acting as major access routes to qualifications for staff who otherwise would tend to remain stranded at the lower levels of organisations. Courses like the certificate for wardens of sheltered housing are also important in recognising the quality of work done by a predominantly female group and providing access into the educational system for those who want it.

The IOH, by agreement with the colleges, instituted monitoring statistics with ethnic and gender breakdowns for students. However, the IOH is experiencing difficulties in getting responses from colleges. It does not yet monitor college staffing statistics, though this is being considered.

This small inquiry has illustrated the process of implementation of equal opportunities by the educational bodies, and the way in which they have responded to external and internal pressures. It demonstrates how an institution may have a policy but, until there are some means of monitoring it, the measures may seem ineffective. Effective monitoring is also dependent on sensitivity to equal opportunities issues by those who carry it out, or on the creation of effective criteria and performance indicators.

Local action

Employers, staff and IOH branches clearly have a part to play here by requesting that those statistics which are required nationally be presented to local consultative groups. Employers, in their decisions on what staff to send on courses and how many, exercise a major influence on the local provision of courses. They can also monitor the membership of governing bodies and see whether local publicity reflects an equal opportunities approach. They can also review external provision to see if they can provide more support for women in technical or management jobs.

Table 9.3   Questions at behaviour level

---

6 Do staff reflect an awareness of equal opportunities issues in their behaviour towards other staff, employers and students?

7 Is suitable action taken regarding issues of sexual harassment or discriminatory language involving staff or students?

8 Are men and women encouraged to contribute equally to class discussions and activities? Are issues regarding unequal participation or stereotyping and disadvantage sensitively discussed?

Behaviour

Local action

This is a more difficult level to monitor than the statistics, but one where local knowledge is more likely to be effective than national action. Training officers who visit colleges are one source of information through their interaction with staff. Sensitivity to feedback from students, especially those who drop out, can also be important. This feedback needs checking against other sources, otherwise there may be unacceptable pressure on the individual involved. Similarly, if complaints about discrimination or sexual harassment have to be made, evidence can be gained from the way the complaint is handled, as well as its content.

Objective research methods can be used, for example, to monitor classroom interaction or language, and are well within the scope of student projects. If necessary, a student on an entirely different course could do this. If employers spell out their own policies and procedures, they have better standards with which to assess colleges. It is also important that the same standards are applied to housing employers' internal and contract training.

Content

Although individual institutions began to review teaching content and practice in the early 1980s, progress was by no means uniform. It was difficult for individual courses to overcome male staff resistance until there was full-scale institutional backing. In many institutions this came by the mid- or late 1980s. Modern course guidelines tend to give less specific guidance about content, and educators need to work hard and think through the issues.

Table 9.4   Questions at content level

---

9 Does the content of teaching and assessment include sufficient development of understanding of the ways in which class, gender, race and disability affect life-chances and, in particular, access and needs for housing?

10 Does the content of teaching and assessment include sufficient development of understanding of the processes of stereotyping and ways of improving equality of opportunity?

11 Is this understanding of the processes of discrimination then integrated with and applied to other areas of the curriculum?

There is some good material; the major problem is in finding enough time and getting it integrated into all aspects of the course.

The IOH currently pays attention to this issue in its validating process. But it is at this level that local monitoring by housing employers and staff can play a crucial role not easily met by national organisations. Employers can often have access, not only to the formal syllabus or guidelines, but also to teaching schemes, assignments and marked examples, so they have a mass of information with which to test the incorporation of equal opportunities into the curriculum.

CONCLUSION

Education and training have gone some way towards promoting equal opportunities but much more can be done. It is essential to be vigilant and maintain the gains we have made. But now is a good time to act. Government changes in finance and administration are making colleges and polytechnics subject to market forces, and changes brought in by the NCVQ mean that employers are not going to be quite so tied to local college provision. It is surely time for housing organisations to use their power as buyers to influence the institutions – as well as influence flowing in the other direction. The practical checklist has already been tried out at national and local levels as a guide to employers and staff in the housing service.

NOTE

1 The Institute of Housing became the Chartered Institute of Housing in 1993.

REFERENCES

Brion, M. (forthcoming) *Women in the Housing Service*, London: Routledge.

Brion, M. and Tinker, A. (1980) *Women in Housing*, London: Housing Centre Trust.

Byrne, E. (1978) *Women in Education*, London: Tavistock.

Connell, R.W. (1987) *Gender and Power*, Cambridge: Polity Press/ Blackwell.

Crook, T. (1985) *Strengthening Committees*, London: NFHA.

DOE (1990) *Training, Education and Performance in Housing Management. Efficiency Report and Action Plan*, London: DOE.

Education and Training for Housing Work Project (1977) *Housing Staff*. London: The City University.

Garrett, S. (1987) *Gender*, London: Tavistock.

Greed, C. (1991) *Surveying Sisters*, London: Routledge.

Griffiths, S. (1991) 'His and not her story', *Times Higher Education Supplement*, 7 June.

Hakim, C. (1979) *Occupational Segregation*, London: Department of Employment.

Kearns, A. (1991) *Voluntarism, Management and Accountability*, Glasgow: Centre for Housing Research.

Lakoff, R. (1975) *Language and Woman's Place*, New York: Harper & Row.

Levison, D. and Atkins, J. (1986) *The Key to Equality*, London: Institute of Housing, Women in Housing Working Party.

Luthra, M. and Oakley, R. (1991) *Combatting Racism through Training: A Review of Approaches to Race Training in Organisations*, Coventry: University of Warwick, Centre for Research in Ethnic Relations.

McGivney, V. (1992) 'Women and vocational training' *Adults Learning*, Vol. 3: 10, June.

Morris, J. (1988) *Freedom to Lose: Housing and People with Disabilities*, London: Shelter.

NFHA (1985) *Women in Housing Employment*, London: NFHA.

NFHA Women's Standing Group (1992) *1992/3 Work Programme*, London: NFHA.

Rao, N. (1990) *Black Women in Public Housing*, London: London Race and Housing Research Unit, Black Women in Housing Group.

Smith, G. and Bailey, V. (1993) *Staying the Course*, London: BTEC.

Spender, D. (1980) *Man Made Language*, London: Routledge & Kegan Paul.

Spender, D. (1982) *Invisible Women: the Schooling Scandal*, London: Writers and Readers Publishing Cooperative Ltd.

Waring, M. (1989) *If Women Counted*, London: Macmillan.

Whylde, J. (1983) *Sexism in the Secondary Curriculum*. London: Harper & Row.

Wolpe, A.M. (1977) *Some Processes in Sexist Education*, London: WRRC Publications.

Wolfe, L.R. (1991) *Women, Work and School*, San Francisco: Westview Press.

Chapter 10

Women achievers in housing
The career paths of women chief housing officers

Veronica Coatham and Janet Hale

This chapter examines the career paths of women chief housing officers, employed by local authorities in England and Wales, who were in post in the latter half of 1989. At this time ten were identified, representing 3 per cent of all those holding such a post. The research was undertaken to complement that of Janice Morphet (1993), who investigated the career pattern of women chief executives, of whom only four were identified as holding such a post in England in 1989. The purpose of both studies was to examine the means by which the women achieved these high status posts and to compare their progress with that of other women achievers. Whilst Morphet's study found two clear pathways by which women chief executives achieved their position, there were several routes adopted by women chief housing officers, the main route being through estate management, starting out as housing management trainees. Both the women chief executives and the chief housing officers stated that they encountered no particular problems relating to their gender but all expressed clear views on their management styles. Morphet (1993) comments that this self-knowledge may explain, in part, their success. The results of both studies provide some indication of what women may need to do if they wish to reach top jobs in local government.

BACKGROUND

There are 296 local authorities in England and Wales involved in the provision of social housing, whether as providers or enablers. Well over 50 per cent of the recipients of their services will be women, yet at the time of our research there were only ten women heading a local authority housing service and answerable directly to the Chief Executive. The research also identified a number of women at third-tier level who were in effect responsible for the housing service but, because they were answerable to a chief officer, responsible for a number of functions, say, health and housing, and not the Chief Executive, they were not included.

Two of the ten women holding the post of chief housing officer in England and Wales declined to take part because of work commitments. Interestingly, all those women who did agree to take part were chief officers within district councils and responsible for other functions apart from housing (mainly environmental health, but technical and legal services were also represented). In addition, we were aware that, at the time of our research, there were a small number of women chief housing officers working in Scotland, and also that there were a growing number of women appointed to the post of Housing Association Chief Executive. However, an analysis of their progress, plus that of women at third tier may have to wait for a larger, more extensive study.

INTRODUCTION

The objective of this research was to identify the different routes taken by women to get to the top of the housing profession in order to:

1 provide examples to other women of how they might reach senior positions in housing organisations, and
2 provide information to local authorities on how they might open up opportunities for women.

However, before considering the different routes taken by women chief housing officers, a brief review of the growing body of literature on women in top jobs, both in the private sector and local government, will be undertaken. This will be followed by an analysis of the role of the local authority chief housing officer and its impact on women. Then, penultimately, a more detailed analysis of the research undertaken on women in top jobs is provided as this identifies the main characteristics of women achievers and any limiting factors which may affect their progress. These characteristics are then applied to the women chief housing officers interviewed, their career patterns are examined and issues relating to their gender; in particular, their management style and relations with other staff and external agencies are explored. It should be noted that this was not the main purpose of the study, but it provides an indication of the approaches they adopted to get to the top.

REVIEW OF THE LITERATURE

It is important to consider the current literature on women in top jobs as it helps to identify those general or specific characteristics of women achievers in local government, a sector where there is a clear hierarchy of job progression and promotion is achieved by competitive interview. Hence it can be argued that the structure is clearly provided and ground rules for promotion known. Local government officers are aware of the factors which influence career progression, for in local government the trend seems to be that of moving between local authorities to gain promotion, especially in the early years of a career.

This pattern of progression has clear implications for women; Coyle (1989) maintains that, because the structure of local authority employment is cumbersome and rigidly hierarchical, opportunities for individual women to create change for themselves through job mobility are restricted. She argues that the boundaries between management, professional, administrative and clerical work are formidable, and the opportunity, for example, to move from secretarial to administrative work is rare because of the strict maintenance of pay differentials. This formalised structure, she says, 'hugely disadvantages women, most of whom are not employed in jobs which are part of any recognised career structure' (Coyle, 1989: 42); for example, women working as home-help supervisors, wardens of sheltered homes. Moreover, Coyle comments that, even for those women who are employed in work which can provide a set career path, such as estate management, there is rarely any special provision for re-entry into employment after a career break, and, if there is, it is usually at a lower grade than the post previously left. There have been some notable exceptions, but they are notable because they are so few.

Key research recently undertaken in this area is that of Morphet (1993) on chief executives, Greed's (1991) work on the surveying profession, the LGORU report on women in local government (1981), and more recently the LGMB publication *Breaking Down the Barriers* (1991) which contained an analysis of an extensive survey of over 1000 local authority women managers. *Breaking*

Through the Glass Ceiling by Abdela (1991) focused on providing more opportunities for women in local authorities, and there was also the Hansard Report (1990) entitled *Women at the Top* which examined the roles and positions of women in public life, and provides a good practice guide for organisations seeking to create more equal opportunities for women.

The literature is also of assistance to those officers responsible for recruitment and promotion in organisations where it may be assumed that women fail to progress for reasons unrelated to their work, i.e. because of their domestic roles. There is a further area to be considered, that of women's perceived lack of confidence and attitudes towards the non-functional areas of their work; this will be discussed later in the chapter. The publications of the Equal Opportunities Commission are useful to those who may wish to explore the issue further. In addition, of particular interest may be the SATL publication *A Generic Model for a Computer Assisted Guidance System for Women Returning to Paid Work* (1990) and *Breaking Through the Glass Ceiling*, by Abdela (1991) which is a thorough assessment of local authority initiatives and examples from outside the public sector. Similarly, Rajan and van Eupen's *Good Practices in the Employment of Women Returners* (1990) is a useful source of best practice in relation to women returning to work after a career break, although many of the initiatives documented by the authors have wider potential applications. Specifically in relation to housing, there has been a small but influential number of publications commencing with Brion and Tinker's work in 1980, which was later followed by a report of the National Federation of Housing Associations in 1985, the Institute of Housing in 1986 and the GLC, also in 1986. All these publications discussed not only the position of women as employees of housing organisations but also as recipients of services, with a different emphasis in each publication. Needless to say, all the publications found that the number of women employed in housing was considerable, especially in the lower, clerical and administrative posts, but as the career ladder was climbed (see Figure 9.1, p. 175), so the number of women decreased. Various hypotheses consistent with those we have identified, were put forward as to why this was the case. This will be discussed later.

The appointment of women to two chief housing officer posts in large local authorities in 1990 rekindled interest in the numbers of women in senior positions (Miller, 1990). So there is increasing interest in the growing number of women at the top – but why, when it was women who were the pioneers of the housing management profession?

THE DEVELOPING ROLE OF LOCAL AUTHORITY CHIEF HOUSING OFFICERS AND ITS IMPACT ON WOMEN

The structure of local government in England and Wales comprises metropolitan and district councils totalling 296 in number. The features of each council area can vary greatly in terms of geographical size, population, density, economic infrastructure, etc., with the result that, in the housing service, the chief officer can be responsible for managing stock of 100,000 properties in a large metropolitan area or 2000 in a more rurally based district council, where the chief officer will probably be responsible for more than one function. Interestingly, all the women who took part in the research were chief officers of district councils, with an average of 10,000 properties under their remit, and responsible for more than one local authority function. These local authorities were contained within the 29 per cent of those identified by Levison and Atkins (1986) where the housing function was combined with some other. What, therefore, is the role of the chief housing officer? The role is broad and diverse, and Cantle (in Davies, 1992) argues that the new-style role of the chief housing officer includes the responsibilities outlined in Table 10.1.

This new-style role contrasts interestingly with that of the traditional housing manager, dating from the 1920s when housing management was seen purely as an administrative function. For example, Cantle (1992) argues that this style of management was distinguished by a workload dominated by functional and routine responsibilities, a concern with resolving problems about the provision of services, maintaining control of resources and decisions, and technical knowledge. In recent years a more strategic 'enabling' role has been adopted to ensure provision of adequate housing in an area. This trend has emerged from such political and other developments as Compulsory Competitive Tendering, 'Tenant's Choice' and similar changes to the financial regimes which govern housing. Cantle states: 'the desire to introduce a more open and democratic style of local government with a more sensitive service delivery, and to alleviate discrimination and dis-

Table 10.1 Principal characteristics of old- and new-style directors of housing

Old style	New style
Workload dominated by functional and routine responsibilities	Free of functional and routine responsibilities
Producer oriented: concerned with problems of providing service	Customer oriented: concerned with problems of receiving service
Reacts to problems as they arise	Anticipates, plans ahead and sets objectives
Maintains control of resources and decisions	Attempts to devolve power and budgets
Concerned with equality of treatment and uniformity	Promotes special programmes to combat inequality and disadvantage
Technical knowledge and administrative skills emphasised: professional first	Emphasises political awareness, management and communication skills: manager first
Negotiates with staff and consumer groups	Participative management style and encourages active involvement of consumer groups
Reviews progress when required to do so	Monitors progress against targets with developed performance indicators
Concerned with (limited range of) responsibilities and services provided	Concerned with (comprehensive) responsibilities and services provided, together with work of other agencies, by reference to research-based housing strategy

Source: Cantle, 1992: 23

advantage, requires an entirely new approach and style' (Cantle, 1992: 10).

When questioned about their role as chief housing officer, the women we interviewed perceived it as being a mixture of those elements identified by Cantle as new- and old-style management. However, all identified one of their key roles as being forward planning and coordination, and one specifically mentioned the need to develop an organisational culture and vision. We intend to provide only a brief comment on the development of housing management as a profession as it is so well documented elsewhere (Smith, 1989; Brion and Tinker, 1980; Cantle, 1992), but it is worth noting that housing as a professional area has had to fight for recognition. Laffin (1986) argues that housing management has suffered because, organisationally and inter-organisationally, it is much less well defined than other local authority functions. It is well known that housing is one of the new 'professions' where it is possible to rise to senior status without a professional qualification of any kind, a situation which the Chartered Institute of Housing is now urging housing organisations to assist in remedying.

The development of housing management as a profession evolved in the early 1920s and was dominated by women who were members of the Society of Women Housing Managers. They followed the style of housing management promoted by Octavia Hill, which was welfare oriented but within an administrative framework. At the time, there were no separate housing departments with the housing service being the responsibility usually of the town clerk's office. Indeed, by 1935, only 35 per cent of local authorities had a housing manager in post. The continued development of housing as a profession was facilitated by the growth of local government and the establishment of separate housing departments. Brion and Tinker commented:

> London government reorganisation in 1965 and general local government reorganisation in 1974 produced much larger housing organisations with much more powerful and demanding top management jobs. At the same time the hugely increased salaries made them much more attractive to men. Previous work on factors affecting women's careers has shown that sexual stereotyping works against women being awarded such top management jobs. In many cases where several authorities were merged during reorganisation it was the existing male member of staff who became the Director and the existing female chief officer who became Assistant Director or Deputy Housing Manager. The women who survived local government reorganisation as heads of housing departments were either very lucky or very gifted.
>
> (Brion and Tinker, 1980: 103)

However, ten years before local government reorganisation the trend of more men being attracted to housing work because of status and high salaried jobs had already been recognised when the Society of Women Housing Managers merged with the more male-dominated Institute of Housing Administrators to form the Institute of Housing.

The whole shape of housing work is changing yet again, and it is likely that the 1990s will attract a different calibre of person to housing work, as local government in general changes its orientation to a more 'business-like' and 'market-oriented' approach. This will bring more competition for top jobs as different skills are needed. For the Chief Housing Officer this may involve the need to develop a sense of strategy and individual management style to manage the housing function which may vary according to personal characteristics, individual backgrounds and training as well as the characteristics of the local authorities (Morphet, 1993). Cantle (1992) explores this issue further; he argues that professional officers are more willing to limit their responsibilities to particular service areas rather than be responsible for the management of human resources and controlling delegated budgets. He estimates that 80 per cent of a Director of Housing's time can now be spent on management issues, yet he or she may not have received any specific management training. Training in housing is still focused on professional courses and the development of housing skills; only recently has performance monitoring of both the staff and the organisation received any attention, so necessitating the development of a management culture for housing organisations of the future. One of the keys to this development will be the management of change with a Director or Chief Housing Officer responding to uncertainties and creating visions of the future, and at the same time being more aware of political climate and sensitivities. Cantle stresses that a communicative and participative style of management, where staff are involved in the management of the department, is critical.

Cantle's line of argument will be further discussed when we examine management styles. However, we now consider the experiences of the women interviewed, some of whom got to the top in a remarkably short time, displaying a range of talents and attributes along the way. Without exception, they said that they enjoyed the challenge of the job and its changing nature, but this very point could have implications for other women seeking to rise to senior positions within housing organisations. Proven and Williams (1991) argue that, whatever the outcome for current well-established housing organisations, housing management issues remain a crucial part of the service, and there will be a continuing need for professional housing staff. As the housing management function tends to be delivered by staff at lower middle management level, where there is a high percentage of women, so opportunities for continued employment remain, but *above* this level, where women are less well represented, the need for management training is increased, particularly in the light of Cantle's comments on the future of the housing service. It is notable that none of the women we interviewed had received any formal management training.

Let us now see how women achievers have reached their positions and the strategies they have developed, and how this relates to the successful eight women chief housing officers interviewed for this study.

STRATEGIES FOR THE SUCCESSFUL CAREER DEVELOPMENT OF WOMEN ACHIEVERS

In the last 10–20 years there has been increased interest in the position of women managers, especially those who have achieved high status in their organisations. Much of the research has been carried out in the private sector, both in the USA and the UK, with only limited investigations of public sector organisations. Of note in the public sector has been the work of the Equal Opportunities Commission which investigated the practices of British Rail and the Post Office in 1986 and 1989, both of which operated an internal promotion policy. Indeed, this too is a phenomenon of local government in the UK, which is why, until more recent years, comparisons with the private sector have been of limited value. One piece of research which did attempt to undertake some broad comparison was that of the Ashridge Management College (1980), which, through a series of case studies, sought to identify the limiting factors affecting the development of women in six different organisations. Of critical importance were:

(i) The career paths that had been established over time in organisations and professions.

The research found that in some organisations the pattern was formalised and explicit while in others it was less formal but well accepted. In more established organisations, e.g. local authorities, the career paths had been determined by men as only a few women had been in senior management positions for long periods. It should be noted here that before the Second World War, because of the operation of a 'marriage bar', women were obliged to retire from their careers on marriage, with the result that the Society of Women Housing Managers was dominated by unmarried women. The Ashridge researchers argued that women tend to follow the same career paths as men but face more hurdles, so making it difficult to get to the top and break through the 'glass-ceiling'; for example, part-time study tends to take place at the start of a career with promotion to supervisory grades on successful completion of examinations in the mid-20s, a time when women may be contemplating having children. Of even more critical importance is the fact that generally after 12–17 years' service, management status can be achieved when employees are in their mid-30s. Spencer and Podmore (1987) noted that while in some professions the critical career years were the late 20s and early 30s, these were also likely to be the years when women would experience their most demanding family commitments. This can therefore present a barrier to women wishing to return to work at the same level after a career break, as there are still comparatively few planned re-entry schemes. It is interesting to note that Larwood and Gattiker (1987) found that the age at which managers attained professional status and line management responsibility was an important factor in determining their later success within an organisation. For women following a 'non-standard' (by male criteria) career path, these stages can be delayed, affecting eventual promotion within the employing organisation (Hennig and Jardim, 1978).

(ii) The attitude of senior executives to women.

The researchers found their influence to be subtle and to have an inhibiting effect on the progress of women who tended to be concentrated in jobs regarded as suitable for them, i.e. direct contact or specialist jobs. Such jobs may not be part of the mainstream career path of the organisation and so, in effect through stereotyping, women may be placed at a disadvantage. More recent research conducted by the Institute of Management (1992) confirmed this finding with the Director General stating that 'men are the prime barrier to women in management' (reported in the *Guardian*, 2 November 1992). Similarly, the LGTB survey *Women Managers in Local Government – Removing the Barriers* (1991) concludes that the way forward involves chief executives and top managers taking charge of setting the style for the local authority and changing the long-held assumptions about management style and the role of women managers.

(iii) Women's own attitudes and behaviour.

These were also seen as important limiting factors in that women express less confidence in their ability to tackle jobs; they tended to describe their weaknesses and remained diffident about their strengths. Women were seen to be conscientious and meticulous, strengths which are awarded less value high up the hierarchy where less structured, flexible modes of working are required. However, some of the women interviewed as part of this research said that this latter type of working was more attractive to them. It may be the case that there have not been enough women in 'top jobs' for this premise to be adequately investigated. Apter (1985) reports that women may allow themselves to be sidetracked from pushing ahead at all costs, valuing relationships (work and personal) above success at any price, and they may therefore undermine their career achievements. She provides the following example:

> A girl who stops playing a game because she sees what is happening to her friendships as she plays the game, is not less mature, less determined, than a boy who plays according to the rules and accepts the contest and its consequences. She will not be as good at winning, however, unless the rules of the game change and we set up a new sense of what it is to win.
>
> (Apter, 1985: 71)

Marshall (1987) makes a similar point, noting that:

> Many women have become dissatisfied with automatic careerism and have realised that it meets part but not all of their need for equality . . . several of the senior managers interviewed put their whole lives in the balance at particular choice points . . . if

such action would serve other, more personally significant needs and goals.

(Marshall, 1987: 24–5)

In the light of these limiting factors, how are women to achieve 'top jobs' in large organisations? Are there any prerequisites for women who wish to achieve senior status? Generally, these factors have been shown by current research to be little different to the strategies adopted by men, i.e. purposeful and well planned (Asplund, 1988), but where women do exercise these strategies, tensions may be created both within the organisation and personally for the women concerned (Morphet, 1993; Silverstone and Ward, 1980). In her study of women local authority chief executives, Morphet (1993) identified a number of factors for the successful career development of women:

- Pre-planning
- Receiving specific support during the formative years
- Finding a sponsor/mentor
- Finding a supportive partner
- Minimising the career break
- Location
- Hard work

Each of these strategies is briefly discussed below and then used to analyse the findings of this research into the career paths of women chief housing officers.

Pre-planning

Career planning has not always been considered to be as important for women as it is for men. As women have a variety of roles available to them, i.e. home maker, mother, carer, they are able to some extent to determine whether they wish to work and their pattern of working. However, the General Household Survey 1991 found a sharp rise in working mothers, especially of children under 5 years old, so reflecting the increasingly important role of women in the workplace even in times of recession; the proportion of mothers of under 5s with a job rose from 25 per cent in 1988 to 43 per cent in 1991. This figure comprises 29 per cent in part-time posts and 13 per cent working full time.

Fogarty *et al.* (1981) argued that women were acquiring career planning skills, but Asplund (1988: 18) found that women who had succeeded 'denied any conscious effort on their own part to achieve their present position. They explained their careers in terms of chance and the luck of the draw'. Indeed, some of the women she interviewed seemed embarrassed to talk about their success. Asplund also found that women who had succeeded had had to expend more effort than a man would have needed to, and had generally taken longer to achieve their positions. Fogarty *et al.* quote one senior administrator who said: 'There may be truth in the contention that women have to prove clearly that they are successful, whereas men are assumed to be successful until they conclusively demonstrate that they are failures' (Fogarty *et al.*, 1981: 44).

However, the choice of a career path depends on the choice of profession; some careers may offer flexibility in working hours or access to jobs in new areas if workers are prepared to be mobile. Fogarty *et al.* (1981) argued that a career outside a profession was more difficult to establish as success was more dependent on achieving a wide range of experiences in the early years, rather than following a recognised path, i.e. employment, part-time study, examinations, promotions (see also Ashridge Management College, 1980).

But even within professions it is not easy for women to get to the top. Morphet (1993) found only four women chief executives in the UK with an average of 12 women chief officers in other professional groupings, e.g. housing, planning. It can be argued that it is more likely to be the case that women in second-tier posts may find subtle barriers to their achieving chief officer status. However, none of the women interviewed as part of this research found this to be the case.

Receiving specific support during the formative years

The attitudes and support received by women in their childhood years has been found to be crucial in instilling a sense of self-esteem (Pahl and Pahl, 1971). Of more importance has been the combined influence of career and family demands, especially where they may have become used to the exercise of power. Morphet (1993) found evidence that parental support for higher education and the attainment of a career was important (Silverstone and Ward, 1980; Miles 1986). Fathers especially were found to be instrumental in guiding and encouraging talented daughters towards adopting a career. This was also one of the major findings of Hennig and Jardim's influential work *The Managerial Woman* (1978).

Finding a sponsor/mentor

Studies such as the Ashridge Management College report (1980) and Fogarty *et al.* (1981) have found the attitude of senior executives instrumental in determining whether women gained promotion. Where senior management were sympathetic to women achieving senior positions it was easier in practice (Fogarty *et al.*, 1981). Fogarty *et al.* also noted that the personal support of patrons or mentors was just as important in the careers of successful men as women managers. However, Asplund (1988: 27) noted that a woman may miss 'the kind of informal and often confidential communication which her male colleagues enjoy with their bosses', and hence receive less support than her male counterpart unless a special relationship is formed with her senior. As yet, because there are only a few women at the top, there are a limited number of role models for women. Clutterbuck and Devine (1987) noted a study by the American Center for Creative Leadership which had shown that having a mentor at executive level was more influential in the careers of women managers than their male counterparts. Positive action will be needed if this proves to be the case in the UK.

Finding a supportive partner

For men this has always been accepted as the case (Pahl and Pahl, 1971) but women too need support, though of a different type; i.e. the support that men receive is usually in the form of domestic help and childcare or management of the home. Women, however, may need support in terms of morale and confidence-boosting, and, as her career develops, her partner may need to become less mobile (Fogarty *et al.*, 1981) and take on a greater domestic role (Cooper and Davidson, 1982). Recent studies, however, show that women still undertake the larger share of domestic responsibilities. Moreover, while some studies have shown that successful career women may choose to remain single, and not remarry after divorce, Asplund (1988) found that it was more common for career women to marry and have children, yet divorce was still frequent among women pursuing careers. She comments that whereas women used to be either 'single' or have few children (Fogarty *et al.*, 1981; Hennig and Jardim, 1978), a new generation of career women is emerging who are likely to have families but be more prone to divorce because of the inherent stress on the relationship. A woman may also grow away from her partner or 'develop out of the relationship such has been the value placed on a career' (Asplund, 1988: 63).

Apter (1985) discusses research undertaken by a large American company into the life history of its senior female executives. Although 95 per cent of male executives were married at the time of the research, only 47 per cent of women executives were (28 per cent had never married, 4 per cent were widowed and 21 per cent were divorced). Apter comments: 'Those who were divorced or separated spoke of their career as an important factor in the breakdown of their marriage' (Apter, 1985: 140).

Minimising the career break

For women, minimising the career break, combined with raising young children, has always been a critical factor. 'Pauses may delay a woman's career and may even stop her having a career' (Asplund, 1988: 25). McKee (1988), in her study of the legal profession, found that the issue was clear-cut – 'a choice had to be made' at a time when men were able to progress.

It can be argued that there is no ideal point in a woman's career when it is best to have children, regardless of the stage that a career has reached. Childbirth and child-rearing do not necessarily mean the end of that career; however, what it does mean for women returning to work is that they may do so under a new set of conditions (Brannen, 1987).

It is inevitable that the responsibilities of motherhood will have numerous effects on a woman's life, both in employment and personal life. Brannen argues that women develop strategies in order to balance the many competing demands on their time, energy and affection. The ease of transition into a new form of working relies, to some extent, on an employer's attitude towards women in the organisation, e.g. some banks have taken the lead in making provisions for women to return to work and reducing the effects of a career break (flexible working, nursery provision).

Brannen studied two groups of women: those who intended to return to work after childbirth, and those who were intending to stay at home with their children. She found that women with high status jobs tended to give birth at an older age than those with lower status jobs, but had been working for the same length of time because they had spent longer in the education and training sector before beginning their careers.

Local authority housing departments have tended to be slow in developing policies and provisions for women. Few have workplace nurseries but most have developed maternity packages to encourage women to return. However, provision is no better than in the private sector where there are only a few notable examples, e.g. BP, Midland Bank, etc. The continual publicity given to women bringing cases against their employers for unfair dismissal following maternity leave shows that this issue is still critical for women. If women are denied the opportunity to return to work after the birth of their children or face rigid employment practices, how are they to take a positive attitude towards career development and themselves? Abdela (1991) explores a number of useful ways in which local authorities and private sector organisations are attempting to address these issues, and Rajan and van Eupen (1990) show that there are a number of public and private sector organisations taking positive steps to encourage women to return to employment after career breaks. These steps could include a review of accepted working practices through job redesign, if women returning to work with caring responsibilities are able to maximise their potential.

Location

Proximity to an urban centre is one of the key factors in determining whether a woman can pursue a career, especially if she has a young family. Wider choices of career are available which, with a well-developed urban infrastructure, make it easier for women to work.

The other critical factor relating to location is the effect of job moves when there are two careers to consider. It is difficult to move two careers at the same time as a whole variety of domestic arrangements may need to be established. Clearly, it is easier for single people or one-parent families to move; but there are rare cases where one partner is prepared to sacrifice his or her career for the benefit of the other.

Hard work

Why should it be surprising that women fail to reach the top where they will be required to work 12–14 hour days, have little time off, face long and tiring journeys and make sacrifices in their personal life? Asplund comments that even childless women may be unprepared for this situation, let alone those with dependent children. For those who are prepared to face the rigours of life at the top, women often had to work harder than their male counterparts (Miles, 1986; Cooper and Davidson, 1982; Hertz, 1986; Asplund, 1988). For those with young children the effort will have to be greater. There are, therefore, a number of factors which can influence whether a woman can succeed in her chosen career. No one factor may dominate, indeed, the circumstances of each individual may differ. However, for those women who do achieve, there are certain common factors, such as those discussed above, which have been used as the framework for charting the career paths of women chief housing officers.

THE CAREER PATHS OF WOMEN CHIEF HOUSING OFFICERS

In order to identify those women who were chief housing officers a trawl was made of the available literature. To ensure that the information collected was up to date, a letter was placed by the authors in *Inside Housing*, a weekly housing publication. For those women chief housing officers who agreed to take part in the research, eight out of a total of ten, a structured interview was carried out and a review of their job description and CV was undertaken. All this activity took place during 1989 and 1990.

Pre-planning

The majority of the women chief housing officers did not plan their careers, though it was evident that at some early stage in their careers they took control, having realised that they had the potential and determination to succeed. There was one notable exception who said that, when she left university and was contemplating a future career, she identified a number of criteria which her chosen career should have. She then applied for a number of traineeships using these criteria, and eventually accepted a housing traineeship because housing work and the training provided met all her requirements. She took a little longer than the average of $14\frac{1}{2}$ years to become a chief officer.

While the majority of the women interviewed stated that they did not start out with a career plan, they eventually chose housing work as their career as it offered potential for progression and variety of work. Indeed, four of the women started their working lives as housing management trainees, which meant that they had the opportunity to study for the professional examination of the Institute of Housing and at the same time follow a structured training programme provided by the employing organisation. It was clear that all the women took advantage of new and different experiences as they arose; for example, only two women chief housing officers stayed in one post longer than three years, with the average length of time in any one post (excluding the present) being two and a half years. Fairly early on, therefore, all the women adopted a logical career progression, with estate management being the most dominant area of work. However, two of the women had worked in more specialised area renewal, and another two, unusually, on the technical side in surveying and repairs. One of the women referred to her career progression as being the result of 'careful and lucky planning'. The number of posts held before that of chief housing officer ranged from four to seven, and the number of authorities worked for averaged between four and five.

All the women were in their late 30s or early 40s, the average age being 41. However, the youngest woman chief housing officer, on appointment, was 31. Most of the women had been appointed two or three years before the interview, with the exception of one who had been in post for eight years. All had been successful in obtaining a chief officer post fairly early on in their careers, with the shortest period being eight years and the longest 23 years. The latter woman, however, was from a non-housing background. Interestingly, the chief officer who started her career far later than the others was the one who achieved chief housing officer status in eight years! Four of the women were appointed to the post of chief housing officer at their first attempt, but the other four had applications in for several similar jobs at the same time. Therefore, all the women had made a definite decision to achieve chief housing officer status, a situation confirmed by talking to them. Moreover, four were promoted from within their organisations following a short period at second-tier level. Several mentioned that they were usually the sole woman interviewed as they progressed to higher level jobs.

All the women interviewed were chief officers of non-metropolitan housing authorities which did not operate equal opportunities procedures. This reflects the findings of the Institute of Housing's survey carried out in the mid-1980s. All the women headed up joint departments, usually housing and health, with the exception of one who was also responsible for legal services. It is worth noting that two of the eight women came from non-housing backgrounds but both, when interviewed, stated that housing composed the bulk of their work, because of the diversity of functions covered, the number of staff for whom they were responsible and the political involvement of councillors. They were not alone in expressing enjoyment in their work, as will be discussed later.

The achievement of the women interviewed has been remarkable but, owing to the lack of comparative research into the career paths of men who have also achieved chief officer status, it is not possible to say with any conviction whether their achievements reflect the norm or whether women have, on average, achieved chief officer status in a shorter space of time. Certainly, this research supports that of Morphet (1993) who also found that women chief executives had spent only two or three years in each post, including a significant period at second- or third-tier level.

Receiving family support in childhood

All the women interviewed mentioned the strong influence of their families, and in particular the encouragement and support of their parents, both at school and later in making decisions which affected their further studies and future careers. While the father was mentioned by several as being the key figure, others spoke of the encouragement that their mothers and grandmothers gave them. The latter had experienced frustration at not having the opportunity to follow careers themselves and wished their daughters to have a different experience. All but two of the women chief housing officers went to university and obtained a first degree, though two went as mature students. Indeed, four of the women left school at 16 and obtained jobs either as trainees in their chosen profession or as clerical assistants. Most described their upbringing as working class, making their success even more noteworthy as all had obtained a professional qualification of some sort early in their careers.

Finally, on this point, all but two of the women were the eldest in their family, with the remaining two being the youngest of daughters who had all received the support of their parents. This therefore reflects the more general findings about successful women, as already discussed, in that they were accustomed to the exercise of power in the home.

Finding a sponsor/mentor

Only those women who had followed the traditional housing route to the top mentioned mentors, who were without exception men and the chief officers of the organisation in which they were working. This is an important factor and was explored at length in the Ashridge Management College report (1980), which found the attitude of the chief executive to be critical in influencing the progression of women in their organisation. The mentors identified by the women were all spoken of warmly for providing support and encouragement at certain stages in their career development and for the close professional working relationship which ensued. Other women spoke of the positive support received from school teachers and work colleagues, but two mentioned working for members of the Society of Women Housing Managers, which had provided unforgettable experiences. While one of these women was described as 'terrifying', she none the less provided encouragement to the woman concerned in the early stages of her career, which was acknowledged gratefully. None of the women had received further education and training opportunities as they progressed in their organisations, which is a factor that needs to be noted by housing authorities if more women are to succeed.

Dawson (1993: 21) notes that not only does mentoring provide 'the opportunity for people to communicate at different levels, so that institutional knowledge and skills can be passed on directly', but also that the need to learn continues even when managers reach a senior position, 'a point sometimes missed in management development programmes'. Indeed, one respondent to Clutterbuck and Devine's (1987) survey of women managers and women entrepreneurs went so far as to comment: 'Even though I have reached a high level, I feel I now need a mentor in an even more senior position. I used to have several mentors, but I am the same level as them now. Is this perhaps why many businesswomen cannot easily get over the last hurdle?' (Clutterbuck and Devine, 1987: 107).

Finding a supportive partner

All the women had had, or still have, partners. Five had married very early on in their careers and were still with their partners, but three had been divorced – two twice. Two of these three women were still single at the time of the research but one was about to remarry; the third had a new partner. Most of the women had received considerable support from their partners; indeed, two had partners who were prepared to follow their wives' careers though in one marriage this had caused tensions which were not resolved until the husband was successful himself. Other women had partners also in demanding jobs which placed some constraint on job mobility, as was noted by one woman, who had had to compromise her career progression in the early stages in order to accommodate her partner's career. There was only one instance of a relationship failing owing to resentment of the woman's future earning potential, while another woman stated that she grew out of her relationships despite her husbands' being encouraging and supportive. A 1984 survey by the British Institute of Management found that more women managers experienced a breakdown in their marriages than their male counterparts.

There was considerable discussion about how the women achieved a balance between their professional and personal lives. Most had domestic help, though for varying periods of time, and a couple mentioned the role of their partners and mothers in assisting in the running of the home. One woman clearly found it a struggle to achieve a balance and said she would like help but had never found the time to advertise. As a consequence her leisure time was more limited than that of the others. These findings, with the one exception, reflect those of senior civil servants and other women managers who all tended to have organised domestic arrangements. However, having domestic help did not necessarily mean that the women had achieved a balance between their professional and personal lives (see page 209). For those women with children, careful organisation was needed to ensure that the children were cared for after school, and even those with partners said that they were careful to ensure some time together, even if it meant attending official or job-related functions together. There was no common view on this point as some clearly 'lived' their jobs out of choice.

Minimising the career break

It is significant that only two of the eight women chief housing officers interviewed had had children, and that they had them during their career progression and not before. And in order to minimise the effect on their careers both took a limited amount of maternity leave, returning to work quickly and employing full-time nannies when the children were young.

One of the two women with children was in a permanent relationship, while the other had been divorced, but both had children early in their careers. Some of the women commented that they felt it would not be possible to have children and 'carry on' in their present post, such were the demands of the post, but they had given it serious thought. It appears therefore that, in the field of local authority housing, there are few women prepared to 'trail blaze' to the top with children (Asplund, 1988); perhaps this is more a reflection of the nature of the work and its demands rather than any negative desires on the part of the women. Because only two women had children, this was not a significant factor to be considered, though the lack of children is.

Location

The research provided some interesting findings as there was no clear pattern. Up to three of the women had moved their geo-graphical location in order to obtain promotion, with one moving for personal reasons. The remainder obtained jobs within travel-ling distance (one travelling up to three hours daily). Certainly, those women on the fringes of large metropolitan areas had a wider choice than those in a more rural location. As well as proximity to work, a key factor in applying for promotion for most of the women was the implication for their partner's career, though as has already been noted, two partners had elected to support their wives' careers and move with them.

Hard work

Because of the nature of their position as local authority chief officers, the women interviewed were obliged to work long hours, on average a ten-hour day, and attend evening and weekend meetings and functions, etc. This pattern of working was something which they accepted, but they all made sure that they 'programmed in' some leisure time – everyone had developed individual strategies for coping. Some took work home regularly, others did when necessary and one or two not at all, but this often necessitated working longer hours in the office. A few described themselves as 'workaholics', though this was not something they all relished. One or two commented that they had to be very strict to ensure that work did not intrude too much on their domestic lives.

WOMEN CHIEF HOUSING OFFICERS' MANAGEMENT STYLE: IS IT AN INDICATION OF THEIR SUCCESS?

The women chief housing officers whom we interviewed, without exception, left us with lasting impressions of their confidence, determination, competence, talents and personalities. They were well qualified for the job and had no difficulties in getting to the top. They tended to be the sole woman on the corporate management team (and sometimes the youngest person), and often the only woman on their internal management team, a situation none expressed any problems with. The problems which did emerge, though, tended to involve employees further down the organisation, who found the appointment of a woman to the post of chief housing officer 'too much of a culture shock', or men representing external agencies who, on occasion, did not take them seriously. This behaviour, they all agreed, was the exception rather than the rule. None talked of any 'real' disadvantages. One or two women discussed their management style and said they had adopted an 'open door policy' which, while maintaining good staff relations, meant that they did not always get through their work in the office, resulting in their working at home. Similarly, others talked of delegating responsibility and empowering staff and the need for a strong internal management team.

   There is a body of literature exploring male and female management styles and, although this was not a major focus of our research, it is worthwhile including some of the major themes at this point. The issue of management style cannot be wholly separated from that of organisational culture, as the latter may influence 'acceptable' styles of management. In particular, Marshall (in Clutterbuck and Devine, 1987: 14) notes that 'Working in predominantly male environments, women are encouraged . . . to copy men's styles of behaviour, attitudes and career aspirations', and Mueller (1984) states that 'It is the male style which is dominant in any large organisation . . . this may lead to bias' (cited by Walters, 1987).

Not surprisingly, the first wave of women senior managers in the 1960s and 1970s used authoritarian male management styles, but it can be argued that there is no one 'right' management style, although some may sit more comfortably within an existing organisational structure. The second and subsequent waves of women managers are using more female management styles, rather than feeling that they must replicate the male model (Rosener, 1990). The female management style focuses more on power-sharing, disseminating information, encouraging participation, developing the skills of junior staff, etc. There is an inherent danger in dividing management styles by gender, in that managers may need to employ a mixture of male and female management styles, ensuring that they are aware of their personal gender-related qualities to 'blend together the wisdom and strength from within offered by female grounding and the outer layer of protection provided by use of the male principle' (Marshall, 1987: 27).

It may be more appropriate to consider management styles in ways other than gender based. Vinnicombe (cited in Beck and Steel, 1989) established in her research that significantly more women are catalysts and visionaries using the Myers Briggs Type Indicator (MBTI) which classifies people according to four Jungian dimensions. Visionaries are seen as natural strategic managers, while catalysts work well with all types of people; in contrast, 56.9 per cent of male managers are traditionalists.

While some research has suggested that there are differences in the management styles of male and female managers, our perception was that the points mentioned above could be equally applied to progressive managers of either gender. Dawson (1993: 21) comments that: 'arguments about . . . gender differences build upon traditional stereotypes and may ultimately be used against women managers and should be avoided.'

However, in order to rise to their positions of power, women may have to be better than men (mentioned by at least two women), more determined and prepared to work the long hours required. Scase and Goffee *inter alia*, comment on the ways in which women may be:

> forced to 'overachieve' in order to obtain senior positions within male-dominated organisations. Some feel . . . that to be successful they have to sacrifice many allegedly 'female' attributes such as 'sensitivity', 'consideration' and 'intuition' in order to develop more impersonal and less affective styles.
>
> (Scase and Goffee, 1989: 153)

For one or two the stresses of working long hours under pressure were evident, but over half said that they did not feel any pressure and were able to dictate the hours they worked; one was studying for an Open University degree.

It is undeniable (Cantle, 1992) that we are entering a further period of change for the housing service, and Cantle concludes that, as the workforce diversifies, traditional management culture and style are giving way to management based on involvement and participation. As Rosener states:

> Fast changing environments . . . play havoc with tradition. Coming up through the ranks and being part of an established network is no longer important. What is important is how you perform. Also, managers in such environments are open to new solutions, new structures, and new ways of leading.
>
> (Rosener, 1990: 125)

While this stage has not yet been reached in the UK, it is possible to look forward to the time when ability is judged on current performance, and then perhaps there will be an end to the male/female management style debate.

CONCLUSION

We have not discussed the impact of equal opportunities to any extent as it did not figure largely in our research. The impact of equal opportunities legislation on local government and housing organisations in particular has been patchy; this is confirmed by

Coyle (1989) who found that practice varies dramatically within organisations which describe themselves as equal opportunities employers. Levison and Atkins (1986) found that examples of good practice, such as the employment of specialist staff to implement positive action training, flexible methods of working, etc., were centred in metropolitan districts and the London boroughs. Because of the variability of equal opportunities practice (see Coyle, 1989), we argue that research into management styles provides more useful information for women seeking senior management positions, and for employers wishing to promote management opportunities for women in their organisations. What conclusions can we draw from this study that may be of use to other women and local authority officers responsible for the recruitment processes and training policies within their organisations?

First, we stress the importance of pre-planning and/or the need for women and organisations to recognise and develop their potential.

Second, there is the need for women to gain wide and varied experience early in their careers, paying special attention to developing their ability to liaise with elected members and senior officers of other departments.

The importance of post-entry educational and training opportunities is critical for all women and this support should be continual as their careers develop. These opportunities were not available to the women interviewed. Employers need to ensure that training schemes genuinely offer equal access to all employees, and are not exclusively evening or weekend based. Dawson (1993) notes that authorities wanting to develop women managers need to think creatively about training opportunities. Family commitments may make residential intensive training and development courses impossible for women to attend. Beck (1990) points out that, while employers may pay for training which will equip a woman to carry out the job for which she has been employed, they may be less willing to fund training which will benefit a future employer, unless they are sure that the employee is looking for a long-term career. Abdela goes further, noting Women and Training's finding that 90 per cent of working women 'do not have access to any suitable training' (Abdela, 1991: 32).

Women also need to take risks and follow up opportunities as they arise, a point emphasised by two of the women interviewed.

Finally, more organisations need actively to consider providing childcare facilities and introducing flexible working patterns. The women we interviewed had made a choice, in the majority of cases, not to have children, which is something they have in common with other women achievers, but why should this pattern continue? Women 'trail-blazers' should be encouraged and supported by the provision of adequate childcare facilities and flexible working arrangements which will allow them to continue with their careers and reverse the under-representation of women at senior levels in local government and in housing organisations in particular. In a 1990 survey commissioned by the Woolwich Building Society, it was discovered that 51 per cent of employees felt that crèche/child-minding assistance would be an important work benefit, and 44 per cent were in favour of job-sharing and term-time working (in Abdela, 1991).

But what of the women chief housing officers themselves? What does the future have in store for them? Most were thinking of the future and planning ahead. One or two talked of moving into consultancy; others of progressing to the post of Chief Executive; some of moving to larger organisations, either a local authority housing department or housing association, with the remainder seeking a 'quieter life'. Two women felt that they were 'ahead of themselves' when they were appointed to the post of chief housing officer and wanted to think about their next move carefully, realising that the housing service was about to undergo a period of rapid change.

Since these interviews took place four women have moved on to other jobs: two out of local government altogether, and two becoming directors of other housing departments. The remaining small group of women have been joined by six more women, two of whom are chief housing officers of London boroughs. A positive trend!

The women interviewed were few in number, and therefore isolated – indeed, some commented that they did not know who the others were, so reaffirming isolation in male-dominated hierarchies. Some aspects of their management styles reflected a mixture of the styles identified on pages 201–4, and at this point it may be useful to reiterate that arguments about gender differences in management styles can be divisive and should be avoided (Dawson, 1993). As more women reach senior positions in housing organisations, so they may begin to impact on the old styles of management identified by Cantle (1992), and, it is to be hoped, change the dominant organisational culture.

REFERENCES

Abdela, L. (1991) *Breaking Through the Glass Ceiling*, London: Metropolitan Authorities Recruitment Agency.

Allatt, P., Keil, T., Bryman, A. and Bytheway, B. (eds) (1987) *Women and the Life Cycle: Transitions and Turning-Points*, London: Macmillan.

Apter, T. (1985) *Why Women Don't Have Wives: Professional Success and Motherhood*, London: Macmillan.

Ashridge Management College (1980) *Employee Potential – Issues in the Development of Women*, London: Institute of Personnel Managers and Manpower Services Commission.

Asplund, G. (1988) *Women Managers: Changing Organisational Culture*, London: John Wiley & Sons.

Beck, J. (1990) 'Financial considerations', in SATL, *A Generic Model for a Computer Assisted Guidance System for Women Returning to Paid Work*, Sheffield: System Applied Technology Ltd.

Beck, J. and Steel, M. (1989) *Beyond the Great Divide: Introducing Equality into the Company*, London: Pitman Press.

Brannen, J. (1987) 'The resumption of employment after childbirth: a turning-point within a life-course perspective', in P. Allatt, T. Keil, A. Bryman and B. Bytheway (eds) *Women and the Life Cycle: Transitions and Turning-Points*, London: Macmillan.

Brion, M. and Tinker, A. (1980) *Women in Housing: Access and Influence*, London: Housing Centre Trust.

Cantle, T. (1992) 'The role of the director of housing', in C. Davies (ed.) *Housing Management: Changing Practice*, London: Macmillan.

Clutterbuck, D. and Devine, M. (eds) (1987) *Businesswoman: Present and Future*, London: Macmillan.

Coe, T. (1992) *The Key to the Men's Club*, London: Institute of Management Books.

Cooper, C. and Davidson, M. (1982) *High Pressure: Working Lives of Women Managers*, London: Fontana.

Coyle, A. (1989) 'The limits of change: local government and equal opportunities for women', *Public Administration*, Vol. 67, Spring: 39–50.

Davies, C. (ed.) (1992) *Housing Management: Changing Practice*, London: Macmillan.

Dawson, A. (1993) 'Holistic approach to equal opps', *Municipal Journal*, 12–18 February: 20–21.

Donnison, D. and Maclennan, D. (eds) (1991) *The Housing Service of the Future*, London: Longman/Institute of Housing.

Fogarty, M., Allen, I. and Walters, P. (1981) *Women in Top Jobs 1968–1979*, London: Heinemann.

Greed, C. (1991) *Surveying Sisters: Women in a Traditional Male Profession*, London: Routledge.

Gutek, B. A. and Larwood, L. (eds) (1987) *Women's Career Development*, London: Sage Publications.

Hansard Society (1990) *Women at the Top*, London: AL Publishing Services.

Hennig, M. and Jardim, A. (1978) *The Managerial Woman*, London: Marion Boyars; also (1979) London: Pan.

Hertz, L. (1986) *The Business Amazons: The Most Successful Women in Business*, London: Methuen.

Laffin, M. (1986) *Professionalism and Policy: The Role of the Professions in Central/Local Government Relationships*, Aldershot: Gower.

Larwood, L. and Gattiker, U. A. (1987) 'A comparison of the career paths used by successful men and women', in B. A. Gutek and L. Larwood (eds) *Women's Career Development*, London: Sage Publications.

Levison, D. and Atkins, J. (1986) *The Key to Equality*, London: Institute of Housing.

Local Government Management Board (1991) *Breaking Down the Barriers*, Luton: LGMB.

Local Government Operations Research Unit (1981) *Women in Local Government: The Neglected Resource*, Reading: LGORU.

Local Government Training Board (1991) *Women Managers in Local Government – Removing the Barriers*, Luton: LGTB.

McKee, V. (1988) 'Get in the right track, lady', *The Times*, 19 September.

Marshall, J. (1987) 'Issues of identity for women managers', in D. Clutterbuck and M. Devine (eds) *Businesswoman: Present and Future*, London: Macmillan.

Miles, R. (1986) *Women and Power*, London: Futura.

Miller, K. (1990) 'New operators on the old boys' network', *Roof*, November/December: 22–5.

Morphet J (1993) *The Role of the Chief Executive in Local Government*, Harlow: Longman.

National Federation of Housing Associations (1985) *Women in Housing Employment*, London: NFHA.

Pahl, J. H. and Pahl, R. E. (1971) *Managers and their Wives*, Harmondsworth: Penguin.

Proven, B. and Williams, P. (1991) 'Joining the professionals? The future of housing staff and their work', in D. Donnison and D. Maclennan (eds) *The Housing Service of the Future*, London: Longman/Institute of Housing.

Rajan, A. and van Eupen, P. (1990) *Good Practices in the Employment of Women Returners*, Brighton: Institute of Manpower Studies, University of Sussex.

Rosener, J. B. (1990) 'Ways women lead', *Harvard Business Review*, November–December: 119–25.

SATL (1990) *A Generic Model for a Computer Assisted Guidance System for Women Returning to Paid Work*, Sheffield: System Applied Technology Ltd.

Scase, R. and Goffee, R. (1989) *Reluctant Managers*, London: Unwin Hyman.

Sear, N. (1980) 'Introduction', in R. Silverstone and A. Ward (eds) *Careers of Professional Women*, London: Croom Helm.

Silverstone, R. and Ward, A. (eds) (1980) *Careers of Professional Women*, London: Croom Helm.

Smith, M. E. H. (1989) *Guide to Housing*, 3rd edition, London: Housing Centre Trust.

Spencer, A. and Podmore, D. (eds) (1987) *In a Man's World: Essays on Women in Male-dominated Professions*, London: Tavistock Publications.

Walters, P. A. (1987) 'Servants of the crown', in A. Spencer and D. Podmore (eds) *In a Man's World: Essays on Women in Male-dominated Professions*, London: Tavistock Publications.

Chapter 11

Questioning the American dream
Recent housing innovations in the United States

Karen A. Franck

The dream of the single family house and neighbourhood is alive and well in the USA in new housing developments, existing communities and local zoning ordinances that protect its continuation. However, this model that idealises and promotes a single way of living and working is also being questioned (Hayden, 1984; Weisman, 1992) and alternatives are being proposed and built (McCamant and Durrett, 1988; Franck and Ahrentzen, 1989; Sprague, 1991; Fromm, 1991; Porcino, 1992; Ahrentzen, 1991). This chapter examines some key features of the American dream and illustrates how these features can be modified in alternative housing that more directly addresses the needs of many contemporary women.

THE SINGLE FAMILY HOUSE AND NEIGHBOURHOOD

The ideal of the single family house and neighbourhood can be characterised by several social and spatial attributes. Four key ones are:

1 that the house be spatially private and self-sufficient so that no interior or exterior spaces or facilities are shared with other households;
2 that the house be inhabited only by members of a single nuclear family, preferably by a wage-earning father, a home-making mother and their young children;
3 that the neighbourhood be composed entirely of free standing houses on similar lots and housing similar nuclear families;

4 that neither the house nor the neighbourhood accommodates wage work, commercial or service activities of any sort.

Not only is this ideal of house and neighbourhood actually realised, since much American housing does meet many of these spatial criteria, but often the ideal is enforced by law. In many communities comprised of single family homes, local zoning ordinances require that houses be detached, that they be located on lots of minimum size, that they be inhabited by no more than a small number of people who are not related to each other by blood or marriage, and that they exclude any wage work, commercial or service land uses (Ritzdorf, 1986).

Clearly, not all housing in the US meets the single family home and neighbourhood ideal, particularly in cities with a large proportion of apartment buildings in mixed-use neighbourhoods. Even such housing, however, approximates the single family home ideal in two respects. First, the social and spatial independence of each unit is of primary importance and, second, the units are designed only for a nuclear family, a couple or a single person and not for other types of household. Although commercial establishments and services may be located nearby, there is still an underlying assumption that such facilities should not be located within the housing itself. Housing that provides for more interaction between households or between unrelated individuals is seen as appropriate only for special needs groups, such as the homeless or the mentally ill, or for people in a particular stage of life that is viewed as marginal, such as college students or the elderly.

The four characteristics I have listed are closely tied to a series of social distinctions and to the spatial separations that embody and enforce those distinctions. Individual households, as well as different types of household and different types of household member, are distinguished from each other and kept separate. Accordingly, each household lives totally separately from every other household. Housing appropriate for young single people, the elderly or couples without children in the form of apartment buildings are often forbidden by law in single family neighbourhoods. Fathers are expected to commute to other areas for employment; mothers are expected to stay home to care for house and family. Individual dwellings, as well as different types of dwelling, are distinguished from each other and kept separate. Thus zoning ordinances define the single family house quite specifically, give it the 'best use' rating, and often forbid any other types of housing within a neighbourhood. Different kinds of activity, particularly those that are deemed domestic and those that are not, are also distinguished and kept separate. So zoning forbids wage work in or near the home and the integration of stores or other services into residential neighbourhoods.

FOR THE GOOD TIMES ONLY

Underlying all these distinctions and separations is a fundamental and powerful desire to accommodate and to *appear* to accommodate the 'good times' only. The house and neighbourhood should shelter and symbolise a life of leisure, comfort and ease that is free of work, illness, old age, and any economic or other strain. The home and neighbourhood should form an idyllic retreat not compromised or contaminated by the presence or apparent presence of work, economic interests, human hardship or human failings. The image of the idyllic retreat requires that the domestic realm should appear to be exclusively a place of leisure even though hard work occurs there.

The house and neighbourhood are envisaged, designed, built and regulated only for the times when all family members are healthy and able-bodied, for the times when they are happy and get along well with each other, and when the household can afford to pay all housing costs from income earned *elsewhere*. Clearly, residents depend on commercial and other services but these must also be located elsewhere. If the house and neighbourhood are to be an idyllic retreat, they cannot accommodate long-term illness or infirmity, an estranged family member, an extended loss of income, or support services for other individual or family problems. Families must resolve these problems or meet these needs on their own without even appearing to have them.

This view of the suburban home and neighbourhood as a kind of vacation from the trials and pressures of daily life comes in part from the origins of suburban design in the imagery of resort communities and cemeteries and from idealisations of the nuclear family and its abode. The guiding idea – for good times only – is apparent not only in the four characteristics listed above but in ordinances, conventions and covenants. Some ordinances restrict all vegetable growing and laundry to backyards only; community covenants may forbid the parking of commercial vehicles, such as the home owner's truck, in driveways. Certainly the concentration on couples with children and no elderly, on the absence of commercial or wage work, service facilities and multi-family housing all serve this desire.

Of course, everyday pressures and life crises *do* take place in any suburban house and neighbourhood. It is not, in actuality, a place of leisure. Homecare, homework, childcare, meal preparation, maintenance, work from the office and other forms of labour all take place. Economic difficulties, illness, strains in the household, spouse and child abuse and other problems occur frequently. Even the four characteristics are not fulfilled: many households are headed by a single female parent and in many households, with one or two parents, the mother is employed outside the home. Many houses are occupied by a single person, by several unrelated people living together or by two- and three-generation families. More and more home owners are renting rooms to boarders and many are building accessory living units in, or attached to, the house.

It is also becoming clear that daily life could only proceed in the suburban house and neighbourhood, historically and in the present, because women spend time, energy and skill bridging the distinctions and separations that are so prized in its design and regulation. Despite their employment, women continue to spend much more time than men in housework, childcare and transporting themselves and their children to school, employment, commercial and service facilities. Their work in the home and in transporting others, although unpaid and therefore unrecognised as 'work', continues to make the house and neighbourhood *appear* to be a place of leisure. Their suffering, individually, each in silence in her own home, of abuse from husbands or lovers continues to make the home and neighbourhood *appear* to be the good life.

Zoning ordinances that forbid wage work (including family day care or elder care), house sharing by unrelated individuals, rentals, accessory units or multi-family housing support the outward appearance of good times only, but increase the burden on women to make life behind the image possible. The ideology is alive and well and so are the actual houses and streets because women work extremely hard to knit together the pieces of daily life that those houses and neighbourhoods sever. The lives of women have changed: many more now carry double or triple responsibilities of home and childcare, formal employment and informal wage earning, but the idealised house and neighbourhood have changed not at all to accommodate them or their families.

All the alternatives presented below modify the American dream by blurring the distinctions or bridging the separations that are so central to the dream. A number of them directly address the more difficult aspects of daily life, modifying the assumption that housing is for the good times only.

CHANGING THE FLOOR PLAN

Modifying the first two characteristics of the single family home and other dwellings that approximate to it requires changes to the conventional floor plan. The latter makes the dwelling unit totally separate from other units; assumes a single, cohesive household epitomised by the ideal of the nuclear family; and assumes interdependent and hierarchical relationships among family members, again as characterised by relationships in a conventional nuclear family of parents and young children. More and more households today do not meet these assumptions; floor plans need to accommodate a greater variety of relationships between and within households.

It is just such variety that is paramount in Katrin Adam's proposal for the conversion of Greenpoint Hospital in Brooklyn, New York to housing for single women of different ages with or without children (see Figures 11.1, 11.2 and 11.3). The original plans for the conversion, sponsored by the National Congress of Neighborhood Women, included: apartments to be shared by two single people; apartments to be shared by two families; swing bedrooms that can stand alone or supplement another apartment; and efficiency units that can be connected to larger apartments. In apartments to be shared by two households, privacy and independence are enhanced by each household having its own door that gives access to that household's private space. The plans thus accommodate privacy and autonomy as well as sharing. This is not the case in conventional apartment house plans where the household is treated as a single unit of interdependent members with some deemed more dependent than others. In Adam's plans a family is seen as potentially needing the support and help of another person, possibly an older relative, who can live in an efficiency unit *and* be part of the family's household.

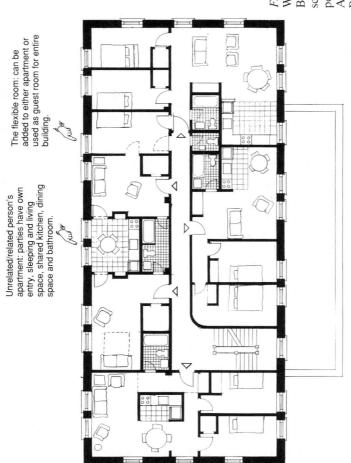

The flexible room: can be added to either apartment or used as guest room for entire building.

Unrelated/related person's apartment: parties have own entry, sleeping and living space, shared kitchen, dining space and bathroom.

Figure 11.1 Neighborhood Women Renaissance Housing in Brooklyn, New York: proposed schematic floor plan with two-person apartment (Katrin Adam Associates, Architects, with Barbara Marks)

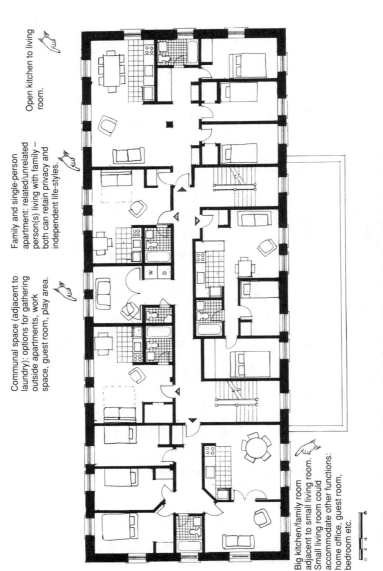

Open kitchen to living room.

Family and single-person apartment: related/unrelated person(s) living with family – both can retain privacy and independent life-styles.

Communal space (adjacent to laundry): options for gathering outside apartments, work space, guest room, play area.

Big kitchen/family room adjacent to small living room. Small living room could accommodate other functions: home office, guest room, bedroom etc.

Figure 11.2 Neighborhood Women Renaissance Housing in Brooklyn, New York: proposed schematic floor plan with swing apartment (Katrin Adam Associates, Architects, with Barbara Marks)

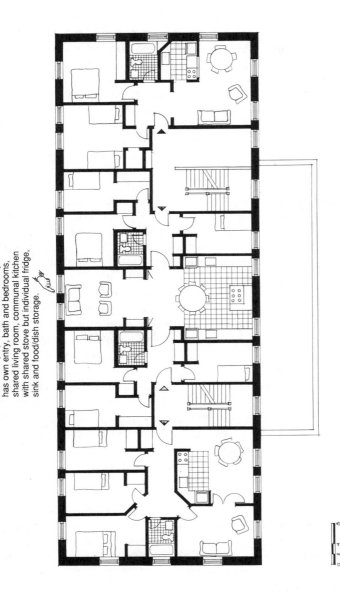

Two-family apartment: each family has own entry, bath and bedrooms, shared living room, communal kitchen with shared stove but individual fridge, sink and food/dish storage.

Figure 11.3 Neighborhood Women Renaissance Housing in Brooklyn, New York: proposed schematic floor plan with two-family apartment (Katrin Adam Associates, Architects, with Barbara Marks)

Hard times are recognised and ways of easing them are built into the plans.

Rental housing in Eugene, Oregon designed for sharing students, single working or retired people uses only a single floor plan that accommodates the needs of four unrelated people (Franck, 1989). These four people share the kitchen and bath/shower; each of the four bedrooms has its own toilet and wash-hand basin. The striking characteristic of the quad is that each bedroom has its own door to an exterior corridor, creating a private entry and exit for each resident. This arrangement gives as much independence and privacy to each resident as possible within the constraints of the shared kitchen and bathroom. Again, the model is not the conventional family with its attendant assumptions of interdependence and cohesion; rather, a household of equal, independent adults. The kitchen, narrow and rather un-inviting in layout, serves a purely utilitarian function. The intentions behind the quads are economic and practical: to give students or other single people housing that has a high degree of privacy and is affordable by virtue of the single-room accommodation, the shared kitchen and the shared shower and bath. The social bene-fits of sharing, such as social interaction, group activities or per-formance of common tasks, may develop among residents, particularly among students who room together for several years, but these are not explicit objectives of the housing.

The small size and narrow configuration of the kitchen in the quad plan does not encourage joint cooking or household gather-ings. Privacy and independence are privileged over gathering and sharing, although both could be accommodated with a different kitchen design. It is precisely opportunities for social interaction and joint cooking and dining that Daniel Soloman wished to provide in the initial design of three shared houses for Innovative Housing in Fairfax, California (Franck, 1989). While the outward appearance of the house models conventional single family homes in image and scale, the interior is designed to house several unrelated people who can choose to cook and eat together in the large common spaces. It is a more socially cohesive model for housing than the quad with more sharing of activities expected and with the independence of each person not as significant. But the equality of the members is, since each bedroom is of equal size.

Both independence and interdependence, privacy and sharing are the objectives of Christine Bevington's proposed plans for

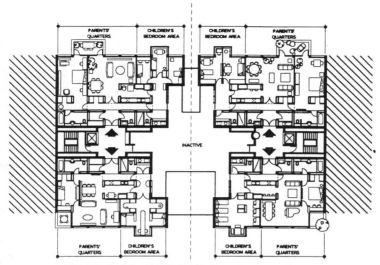

HOME USE DURING FAMILY HOURS

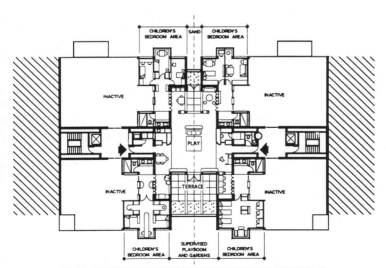

HOME USE DURING WORKING HOURS

Figure 11.4 Four apartments with toggle care (Christine Bevington, Architect)

complete apartments with a common space for childcare. Not only is this space immediately adjacent to the apartments but they open directly on to it. During the day when a caretaker is present, the childcare space becomes a direct extension of part of each of the apartments, with the remaining part of each apartment closed off (see Figure 11.4). Thus, the children and the caretaker have access to the children's bedrooms and bath but the more distant spaces of living room, kitchen and parents' rooms remain inaccessible. Thus a child who wishes to be alone, wants to take a nap or do homework can use his or her own bedroom. Similarly, older children can have privacy while still being in the vicinity of the adult in the childcare space.

Providing this kind of space, for what Bevington calls 'toggle care', preserves the privacy of and offers a high degree of self-sufficiency for each of the dwellings while also building in the ability to share a space and a much needed service. Having this service integrated into the immediate residential setting relieves women of the task of transporting children to and from day care and allows for after-school supervision right in the private home. Bevington's proposal modifies the complete separation and self-sufficiency of each dwelling and, in providing space for day care, bridges the separation between home and support service.

Floor plans can also be modified to accommodate a work space, possibly with its own entrance, or an efficiency unit to provide additional income through rent. Relatively minor modifications to conventional floor plans can accommodate the needs of people or households who do not fit the model of the nuclear family. The modification may be as simple as adding a single door or several doors. This is true in Adam's units for two single people and for the efficiency unit that can be added to a larger apartment or remain separate. It is also true of Bevington's toggle care plan and the quad. Similarly, in incorporating a rental apartment into her own single family house, Anne Gelbspan (1992) gave the unit its own entrance but also left a door between it and the second floor of her own house. This allowed for babysitting by the tenant for Anne's children or by Anne for the tenant's children without either adult leaving her own abode.

INTEGRATING SUPPORT AND SERVICES

The separation of commercial or any other facilities and services from the domestic environment poses difficulties for many women with children, but particularly for single parents who bear all the adult responsibilities in a household. When such women are in a crisis, the isolation of the private dwelling from other dwellings and the trips required to seek support and help from service organisations can make resolution of the crisis or desired change exceedingly difficult.

The recognition that many women in the USA, particularly single women with children, can benefit from support services incorporated into housing has led to the invention and development of many, diverse forms of 'transitional housing'. What all these projects have in common is the recognition that there is a need for housing that is designed and managed for the 'hard times', for the times when women have no place to live, are escaping from domestic violence, are seeking to recover from substance abuse or are pregnant at a very young age. How the project's philosophy, design and programme make these times of transition both possible and easier for women varies tremendously (Sprague, 1991).

Brookview House is a new building developed by Women's Housing Initiatives in the single family neighbourhood of Roxbury in Boston, Massachusetts (Gelbspan, 1991). Although from the outside it appears to be one large single family house, it actually houses seven complete apartments for formerly homeless single mothers and their children. Yet it is also a 'house' with a single, domestic entrance and hall where residents come together as they also do in the yard, the play room and the common living room. It is there that residents attend weekly house meetings and various workshops on parenting, planning a budget and other skills. A full-time manager and assistant manager are present during the day and put residents in touch with other programmes and services. Residents follow a number of rules that forbid drugs or alcohol and overnight guests. The design, the philosophy of the programme and the personalities of the staff all help to create a warm and 'homey' atmosphere, yet the provision of complete apartments also offers privacy and self-sufficiency. Brookview is seven small homes in one larger home. The desire for close relationships between the households and for the unity of the community it houses is clearly manifested and encouraged by the design of a 'house' with the single door and interior shared spaces that a house for one family would have.

Florette Pomeroy House, as part of the Women's Alcoholism Center in San Francisco, fulfils a very different purpose and accordingly adopts a different programme. It serves women with children who are recovering from drug and alcohol abuse and, unlike many such residential programmes, it houses women *and* their children. It is a renovated three-storey Victorian building located next to the Lee Woodward Counselling Center where residents attend counselling and group therapy sessions. Group living is part of the programme and residents are responsible for preparing two meals each day. The residence can house up to nine women and eleven children who stay from six months to one year. Women share bedrooms with each other, separately from the children. This arrangement, which is highly unusual in transitional housing, is intended to help women to develop healthy boundaries between themselves and their children. A staff member reports: 'as part of an entire treatment modality which encourages individuation along with healthy bonding, we put moms and kinds in separate rooms' (Sprague, 1991: 84). The 'homey' quality of the yard at the back and of the community kitchen and dining room with its original wood panelling creates the atmosphere of a warm and welcoming home. Unlike Brookview House, there is little privacy: group living is central to the philosophy, programme and design of Pomeroy House.

All transitional housing modifies several of the characteristics of the single family home. It is intended for single women, with and without children; some degree of sharing of spaces in community spaces or in dwelling units is present; and support services are integrated into the housing – either in the form of staff only or with actual services in the same or an adjacent building. However, unlike other kinds of housing, residents of transitional housing stay only a short time, usually two years or less. The idea is to move from a supportive and interdependent form of living, often with a high degree of supervision and control from project staff, to a more independent, self-sufficient form.

Thus, transitional housing only recognises hard times as temporary, difficulties or crises from which one can recover. What about the hard times that occur on a more regular basis such as a child's or a relative's illness, or those that are permanent conditions like frailty, economic hardship and women's multiple responsibilities for home, family and wage earning? Permanent housing, unless it is for special populations such as the elderly or the mentally ill, does not ease the demands and burdens of daily life. For the most part, permanent housing is still for the good times only.

CREATING NEW KINDS OF COMMUNITIES

The social distinctions and spatial separations of the single family neighbourhood make the lives of women more difficult by discouraging interaction and mutual support between households, by placing places of employment and all services distant from the home, and by forcing a heavy reliance on the private car for transportation. Any alternative models of community must soften or even dissolve the separations and divisions that create these and many other problems (Franck, 1985). A few such alternatives were proposed in the 1980s, notably Dolores Hayden's (1980) redesign of a suburban block and Troy West and Jackie Leavitt's winning entry for the New American House Competition (Leavitt, 1989). Recently, more alternatives have emerged and are being built.

Innovative Housing is a shared housing programme in Mill Valley, California that provides housing, matching services and ongoing support and management for individuals or families who wish to share dwellings. The intent of Innovative Housing is to help residents reduce housing costs, while enabling them to enjoy a higher degree of social interaction and mutual support than they would have living alone. Usually, the organisation rents or buys existing housing for this purpose, but they will soon have a newly constructed group of houses in Fairfax called the Vest Pocket Community. It will consist of five separate houses, each with three or four bedrooms, and an additional building with a community room, a kitchen, a childcare space and a studio apartment. One resident will serve as manager during the evenings and weekends, and a staff member from Innovative Housing will fill this role on weekdays. The houses in Fairfax will not be significantly different in plan or appearance from conventional single family houses, except that all the bedrooms will be of comparable size. The significant difference is in who will live there and in the fact that the houses are grouped together to share additional amenities. While the spatial and architectural differences are relatively minor, the social differences are not.

Cohousing is both socially and spatially different from conventional single family neighbourhoods. Adapted from a Danish model of community, the key goal is a high degree of interaction and interdependence between households (McCamant and Durrett, 1988). All households have their own complete dwellings but share a wide variety of common spaces and facilities indoors and outdoors. Many of these are located in a 'common house' where households share evening meals several times a week. The community is designed and planned by the residents themselves. Offsite parking, a pedestrian-oriented site plan and a range of community spaces are characteristic of cohousing in the USA and Denmark. In both countries residents choose cohousing as a preferred alternative to the single family house and neighbourhood precisely because they seek a higher level of community interaction.

Since the publication of *Cohousing: A Contemporary Approach to Housing Ourselves* (McCamant and Durrett, 1988), the idea has received wide publicity in the USA. Although only a few communities have been built so far (Fromm, 1991), as many as 100 projects are in the planning or construction stages and cohousing newsletters are being published in different parts of the country. According to Charles Durrett, one of the founders of the CoHousing Company which provides information and other services to cohousing groups, the leaders and the most energetic participants in these groups are women and many are single parents who seek the community, safety and shared life that cohousing can offer.

Benicia Waterfront Commons near Berkeley, California is one of the most ambitious cohousing projects to be planned so far (see Figures 11.5 and 11.6). It will provide dwellings for 27 households and a common house of 3500 square feet with a two-storey living and dining room, a kitchen, a children's room, an exercise room, a guest room, a teen room, a laundry, a crafts workshop and a space for pool and Ping-Pong. The site, which is free of cars and partly enclosed by the houses, will include vegetable and flower gardens, a children's playground, a picnic and barbecue area, and 'gathering nodes' or locations where planting, seating or other amenities encourage informal conversations between residents. The units contain one to four bedrooms to accommodate a variety of households with and without children.

Even though cohousing communities in the USA may often retain the detached single family house, they are creating a very

Children's Room

Entry

Guest Room

Bathroom

Down to the rec. room

Kitchen

Bulk Storage

Sitting Room

Dining

Common Terrace

Balcony

0   10'   20'

Figure 11.5 Benicia Waterfront Commons in Benicia, California: common house, preliminary floor plan of upper level (The CoHousing Company and the resident group)

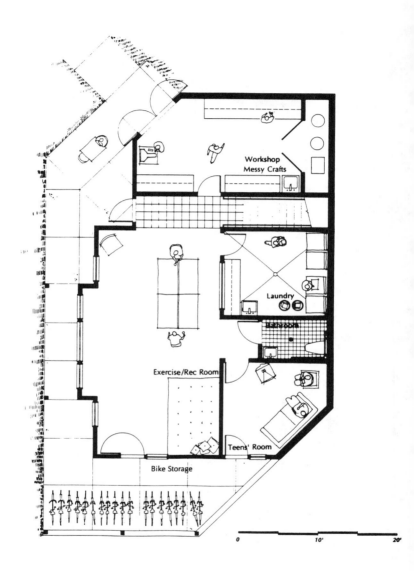

Figure 11.6   Benicia Waterfront Commons in Benicia, California: common house, preliminary floor plan of lower level (The CoHousing Company and the resident group)

different model of community, emphasising interaction between households and sharing daily chores of food shopping, cooking and clearing up after dinner. Rather than hiding this domestic work behind the private façade of each home, cohousing residents make it public by doing it together.

Once the requisite approvals are received, both cohousing and shared housing like the Vest Pocket Community in Fairfax can be built in existing suburban communities. While both types will differ from the surrounding neighbourhoods, they will not alter the overall zoning or fabric of the larger community. This is because they are not set in large-scale communities and because they do not incorporate any commercial or other services. It is at the larger scale of an entire suburban development that Andres Duany and Elizabeth Plater-Zyberk (1989, 1991, 1992) are proposing, and making, changes. They offer a vision of community as a 'town' where types of household and housing are mixed, where commercial and residential land uses are adjacent to each other or actually integrated, where higher density and careful zoning and site planning encourage pedestrian traffic, and where community spaces are of prime importance. So far, the planning by Duany and Plater-Zyberk has been for upper or middle income developments only, but their basic tenets could apply to any population. Moreover, their goal is to create towns that integrate households of different incomes throughout the community (Duany and Plater-Zyberk, 1991). Their work, in theory and in practice, goes further than any other efforts in bridging the myriad social and spatial separations of the suburban ideal. At the same time, it preserves the opportunity to live in a free-standing house inhabited by a single household. Duany and Plater-Zyberk clearly demonstrate that, with careful planning and an innovative form of zoning, preserving this aspect of the American dream does not require the many additional characteristics that divide people and activities.

CONCLUSIONS

The innovations described in this chapter modify several different characteristics of the American dream of house and neighbourhood. In doing so, they address some of the needs of women that are neglected or ignored by the 'good times only' imperative of the American dream. It is crucial that we continue to view, and protect, the dwelling as a private, secure retreat from the public realm, while also exploring various ways of reducing the distance and the division between that retreat and other domains, particularly other households and commercial or other services. Similarly, the dwelling as retreat does not mean that it cannot also serve as a place for earning income either through various kinds of work (Ahrentzen, 1991) or through renting part of it out. It will be useful to address the problem of affordability both through the reduction of costs and through the dwelling's potential to serve in the generation of income.

Imagining, developing and implementing alternatives requires a tremendous amount of patience, determination and hard work. Architects, non-profit organisations and future residents all contribute time and effort to developing housing alternatives when they have the opportunity and are committed to the project. This is very clear among the cohousing groups now working in the USA. Resident involvement is, indeed, an untapped resource. Residents' potential contributions are dramatically demonstrated in Jacqueline Leavitt and Susan Saegert's book *From Abandonment to Hope* (1989) which documents the successful efforts of low income residents in Harlem to renovate and convert their landlord-abandoned buildings to cooperatives. Many of the most active residents were older women.

Even with hard work and determination and a clearly valuable plan, the project may not be completed or may be significantly changed. Although many of the alternatives reviewed here constitute fairly minor modifications to conventional apartments or houses, they were difficult to implement. Katrin Adam's initial plan for transitional housing had to be completely altered in order to meet the formal and informal requirements of the New York City Department of Housing and Preservation. The apartments will now have conventional floor plans. The original plans for the Fairfax Vest Pocket Community were not accepted by the municipality. The final plans meet the requirement that the houses can be inhabited by conventional nuclear families, should that be necessary in the future. A similar requirement was made of the Women's Alcoholism Center. The community planning board required that the counselling centre be built in such a way that it could be converted to housing in the future. In these and other cases, innovation may only be acceptable when it can be transformed easily to traditional arrangements.

The scrutiny or outright hostility of surrounding neighbourhoods and regulating agencies to proposed alternatives indicates the power that the dream of the single family house and neighbourhood continues to exert. This power not only makes it difficult to implement alternatives, it also constrains the envisioning of them in the first place and limits their desirability once they are built. The continuing desirability of the conventional house and neighbourhood in the USA may be part of a nostalgic longing for an idealised version of family life, even though that life no longer exists (Skolnick, 1991) and may never have (Coontz, 1992). Arlene Skolnick warns us: 'Rather than yearning for an elusive perfected family, we would be wiser to consider new social arrangements that fit the kinds of family we now have and the kinds of lives we now lead' (Skolnick, 1991: 224). Considering new arrangements does not mean that we reject altogether the single family detached house in any form. It does mean that we analyse *all* the component parts of the way this house and its neighbourhood have been envisioned, designed and regulated in the USA. We can choose, with great thoughtfulness, which aspects we wish to keep and when. All kinds of variation in social and spatial configurations can be developed. And it does mean that we seek and celebrate a greater variety of options and closer connection between the good times and the hard times.

REFERENCES

Ahrentzen, S. (1991) *Hybrid Housing: A Contemporary Building Type for Multiple Residential and Business Use*, Milwaukee, Wisconsin: Centre for Architectural and Planning Research, University of Wisconsin.

Coontz, S. (1992) *The Way We Never Were*, New York: HarperCollins.

Duany, A. and Plater-Zyberk, E. (1989) 'Kentlands:. town plan', in E. Shoskes (ed.) *The Design Process*, New York: Whitney Library of Design, pp. 226–53.

Duany, A. and Plater-Zyberk, E. (1991) 'A town plan for seaside', in D. Mohney and K. Easterling (eds) *Seaside: Making a Town in America*, New York: Princeton Architectural Press, pp. 86–107.

Duany, A. and Plater-Zyberk, E. (1992) 'The second coming of the American small town', *Wilson Quarterly*, Winter: 19–48.

Franck, K. (1985) 'Social construction of the physical environment: the case of gender', *Sociological Focus*, Vol. 18: 143–60.

Franck, K. (1989) 'Overview of collective and shared housing', in K. Franck and S. Ahrentzen (eds) *New Households, New Housing*, New York: Van Nostrand Reinhold, pp. 3–19.

Franck, K. and Ahrentzen, S. (eds) (1989) *New Households, New Housing*, New York: Van Nostrand Reinhold.

Fromm, D. (1991) *Collaborative Communities*, New York: Van Nostrand Reinhold.

Gelbspan, A. (1991) 'Brookview House: a home for mother and children', in W. Preiser *et al.* (eds) *Design Intervention: Toward a More Humane Architecture*, New York: Van Nostrand Reinhold, pp. 27–50.

Gelbspan, A. (1992) Personal communication.

Hayden, D. (1980) 'What would a non-sexist city be like?', in C. Stimpson *et al.* (eds) *Women and the American City*, Chicago: University of Chicago Press, pp. 167–84.

Hayden, D. (1984) *Redesigning the American Dream*, New York: Norton.

Leavitt, J. (1989) 'Two prototypical designs for single parents', in K. Franck and S. Ahrentzen (eds) *New Households, New Housing*, New York: Van Nostrand Reinhold, pp. 161–86.

Leavitt, J. and Saegart, S. (1989) *From Abandonment to Hope: Community Households in Harlem*, New York: Columbia University Press.

McCamant, K. and Durrett, C. (1988) *Cohousing: A Contemporary Approach to Housing Ourselves*, Berkeley, California: Habitat Press.

Porcino, J. (1992) *Living Longer, Living Better: Adventures in Community Housing for Those in the Second Half of Life*, New York: Crossroads/Continuum.

Ritzdorf, M. (1986) 'Women and the city: land use and zoning issues', *Urban Resources*, Winter: 23–7.

Skolnick, A. (1991) *The Embattled Paradise: The American Family in an Age of Uncertainty*, New York: Basic Books.

Sprague, J. (1991) *More Than Housing: Lifeboats for Women and Children*, Boston: Butterworth Architecture.

Weisman, L. (1992) *Discrimination by Design: A Feminist Critique of the Built Environment*, Champaign-Urbana: University of Illinois Press.

Chapter 12

Innovative housing in the UK and Europe

Tom Woolley

INTRODUCTION

Discussion of housing policy tends to concentrate far too much on questions of quantity as though the solutions to housing need are to be found simply in building more and more houses. This is despite the lessons of the mass housing programmes of the 1960s and 1970s which led to many housing disasters and the demolition of new housing within ten years of it being built. This failure to consider the quality and character of dwellings has been an expensive mistake and yet the quality and design of housing and its suitability for current life-styles are still neglected topics. As a result, we continue to build houses which are unsuitable for the ways in which many people want to live. This will continue to aggravate problems of homelessness as the right kind of housing in the right places will not be available.

There is plenty of evidence that many people today have different ideas about housing than the standard solutions of local authorities and private developers. Women have played a leading role in these changing attitudes and they demand a different concept of the environment from that produced by a largely male-dominated construction industry.

In this chapter, different concepts of housing are described with examples from the UK and Europe of innovative forms of housing that have resulted, in part, from the involvement of the future occupants in the design process. The case for the greater participation of housing users in the design and planning of houses is put forward, as this is the only certain way of ensuring that the housing of the future reflects the needs and aspirations of occupants.

HOUSING DISASTERS

It should no longer be necessary to argue against the dangers and faults of mass housing, after the disasters of the 1960s and the 1970s, and yet the processes underlying the failure of social housing, system building and tower blocks are still largely intact. While architectural styles and fashions may have changed, many of the economic and political forces described in the best analysis of mass housing failures (Dunleavy, 1981) remain largely unchanged. While we would like to think that we have learnt from the mistakes of the past, those in charge of housing policy and action still look for technocratic solutions to our problems. Standards have continued to fall, particularly in relation to space in houses, and the needs and ideas of users and occupants continue to be largely ignored.

The media, political parties and housing organisations are again talking about a shortage of dwellings. As a result, talk of quantity rather than quality is back on the agenda. Construction companies are ready and willing to build large numbers of low standard, poorly designed houses through package deals. These are standardised solutions to housing, much the same as was applied to public housing in the 1960s. Once again, there is a danger that the future occupants of those houses will have little, if any, say in the process of design and construction. An article of mine in *Roof*, Shelter's housing magazine, warning of these dangers was juxtaposed with a full-page advert from Laings, advertising package deal housing (Woolley, 1990).

While the Prince of Wales and a past president of the Royal Institution of British Architects (RIBA), Rod Hackney, and others have used the failure of mass housing to attack modern architecture, they have only done so in stylistic terms. They want to substitute a cosy, postmodern, sentimental style of architecture which is rapidly becoming a new orthodoxy applied by housing associations and other housing providers. However, the real battle against past housing failure continues to be fought by tenants' associations and other community groups. These are frequently led by women who have strong ideas about how they would like to improve their environment which do not necessarily correspond with the new ideas of architects, planners and other experts. Many such tenants' groups have had short-term successes in rehousing, the demolition of tower blocks or the retention of existing viable communities and buildings, and some go further in demanding a say in the planning of new environments. There is much that we can learn from listening to the views of such people and looking at the results of housing that genuinely reflects their ideas.

Housing problems invariably result from the decisions of politicians, housing managers, developers, architects, planners and builders whose attitude to the future occupants of housing ranges from patronising to contemptuous. That the construction industry remains largely male dominated is not unconnected with this, and the current ethos is largely one of little concern and care for the needs of building users.

CONSUMER PRESSURE

However, attitudes are slowly changing, and this is undoubtedly because of pressure from tenants and community groups – the consumers of housing – who expect to have a say in future development. Many architects and others involved in the process have had to learn how to listen and talk to local people and to facilitate the participation of communities in design. This has even received formal recognition from professional bodies such as the RIBA and Institute of Housing (RIBA/IOH, 1988).

Some architects have been able to develop participatory design techniques, trying to train and encourage young architects and other housing professionals to work with people, to sit round a table, use models and easy-to-read drawings, and to organise a process so that local people are involved in design and development (Woolley, 1989b). User participation in design and development does figure in most Institute of Housing courses, though it still receives little attention in schools of architecture. However, in many cases in the UK, where participation actually takes place, it is often tokenistic and manipulative. What is built still reflects the ideas of the politicians and architects far more than those of the occupants. On a European scale, however, design participation has become much more widespread and sophisticated with government-supported facilities, such as full-scale modelling (Woolley, 1987; Dalholm, 1991), and there are now many examples of housing projects where tenants' preferences have clearly been influential. Such successful user-participation projects frequently exhibit many innovative features both in terms of physical design and social organisation. These are features which reflect the genuine needs of the users but are elements often ignored in housing produced without participation.

Many professionals remain sceptical about user participation. They claim that people do not know what they want and that participation makes little difference in practice. Of course, this is often the case because the professionals involved, despite paying lip service to consultation, find it hard to give up their traditionally dominant role and do not make it easy for their clients. It is also assumed that people largely aspire to 'Ideal Home', middle-class suburban ideas of housing when given any sort of choice. Particularly in the UK there is a great deal of pressure for housing to look conventional because of concern about what the neighbours might think or because of worries about re-sale values. Thus, a great deal of participatory housing can look, at first glance, fairly 'conventional' or bear the stamp of the architect involved, but on closer inspection it is possible to discern the influence of users in many different ways. These may be found in the internal layouts, estate planning or the variety of choice available to occupants of house types.

There are now enough examples of participatory housing in many European countries to provide firm evidence that user participation has led to many innovations, not only in the form of housing, but in the way that people live, or want to live. These innovations challenge conventional attitudes to housing design.

Conventional house design and building largely assumes that most people want to live in a conventional semi-detached house with a garden, a garage and occupied by the proverbial nuclear family. Most housing, even where it claims to cater for the special needs of old, single or disabled people, tends to stick closely to these assumptions.

NEW HOUSEHOLDS/NEW IDEAS

On the other hand, some new participatory housing experiments show that many people today want to live in different ways. For example, they want to share facilities, to keep cars out so that children are safe, to live in different kinds of groups from the nuclear family, and in healthy, 'ecological' conditions that do not waste natural resources and energy.

These innovative schemes frequently come from new kinds of households and groups who share communal and progressive ideas. But such groups are not new. There have been many Utopian and innovative experiments in housing during the last 100 years (Pearson, 1988) and these have shared many of the social and political ideals of more recent experiments (Rigby, 1974; Mercer, 1984; Pepper, 1991). While these are often dismissed as the product of 1968 attitudes, radical experiments in living in communes and squatting (Wates and Wolmar, 1980) opened the eyes of many to the possibility of alternative forms of housing and living.

Of course, the organisation, finance and management of new forms of housing are not always easy. Living communally can be difficult and not everyone wishes to do it. Interestingly, many innovative housing schemes have varying levels of communality where different groups have experimented with different approaches to sharing. Also, there are invariably bureaucratic, financial and personal obstacles to be overcome as many rules and regulations tend to reinforce the idea of single family units (Eno and Treanor, 1982). What is exciting for an architect, however, is that these experiments have led to many new and more user-friendly forms of housing design, some of which strongly reflect the ideas and interests of women.

GENDER ASSUMPTIONS

Women who are interested in alternative approaches to housing can learn much from visiting innovative user-participative schemes as they will find ideas that take into account different approaches to living. There seems little doubt that our present environment is produced largely by traditional male attitudes and that there are many gender assumptions implicit in conventional housing (Roberts, 1991). As Roberts (1991: 157) explains: 'The male domination of policy making and design has meant that the presence of a woman is indispensable to a home.' However, she goes on to say that 'While this remains the case, it is difficult to imagine how a gender free environment might be designed' (Roberts 1991: 157).

Whether any of our actions can be free of gender influences is one question, but we can learn much from studying innovative examples of housing that have resulted from user participation. Many of these projects prefigure the kinds of environment that Roberts is seeking. Of course, it might also make a difference if there were more women architects. Currently, only 9 per cent of the UK architectural profession and only 27 per cent of entrants to architectural education are women (RIBA, 1993). However, through user participation, many women have had the first opportunity to describe to architects the kind of environment they would like.

INNOVATIVE HOUSING PROJECTS

There is a growing body of literature on new approaches to housing design. Projects that at one time seemed peripheral and eccentric are now being recognised as central to our knowledge of housing. Colquhoun and Fauset (1991), for instance, devote a chapter of their important book *Housing Design* to 'participatory housing'. They argue that architects need to design small-scale housing developments with users.

> Unless architects get directly to grips with environmental problems of ordinary citizens . . . architects in future will have little or no direct involvement in housing. User participation in design is therefore a new force which architects must harness.
>
> (Colquhoun and Fauset, 1991: 222)

Their book illustrates many recent innovative housing schemes in Europe which show the wide range of imaginative ideas resulting from user participation. These include communal facilities, a wide variety of flexible plan forms, shared external spaces, clusters of housing units and self-building.

One of the principal interests of users, when they have a chance to influence housing design, is that of ecology and health. Many conventional approaches to house building ignore environmental effects and pay little attention to the damaging effects of toxic building materials on the health of occupants. It is not surprising, then, that many so-called 'ecological' housing schemes are a result of design participation exercises. Robert and Brenda Vale, in their book *Green Architecture* (Vale and Vale, 1991), see user participation as one of the fundamental principles of green design.

> If a green architecture seeks to make maximum use of all the resources that are put into the built environment, then human energy and enthusiasm should not be left out. A green architec-

ture should be able to make experts once again the servants of people's needs rather than the masters.

(Vale and Vale, 1991: 136)

To understand the impact of user participation on innovative housing schemes it is necessary to visit them. The rest of this chapter is taken up with some brief descriptions of a few such projects.

Examples of innovative projects in Denmark

One of the earliest examples of innovative housing can be found in Roskilde, Denmark (see Figure 12.1). Here, a group of people got together in 1980 and bought a disused portal frame factory and iron foundry. By using the factory, the group were able to create a large amount of communal space, while providing themselves with low-cost housing units. They added lean-to housing units to the side of the factory building, with each house opening into the communal covered open space. This can be used as a play area for children, for sport, parties or general informal social contact. While space is shared, each family has privacy.

This early scheme, adapting an existing building, became a model for many cooperative social housing schemes, both in the private and subsidised sectors. The idea of a covered circulation or communal space to link houses together makes sense in Denmark's climate. It has been used in many different forms in hundreds of schemes, many of which are illustrated in Franck and Ahrentzen (1991). The use of a covered street linking terraced houses has energy-saving aspects and provides a point of social contact between neighbours, whatever the weather.

One of the principal features of the several hundred innovative collective housing projects in Denmark is the idea of a 'common house'. This is financed by reducing the size or cost of each housing unit and putting the savings towards communal facilities. This was first done in 1973 at Skraplanet, near Copenhagen. Here the communal facilities include a wild area for children, a swimming pool and a community centre, all in a scheme of only 33 houses. Another scheme at Jystrup combines common facilities with an internal street to provide a safer place for children to play, but frequently the common house is in a separate building away from the houses. All the projects have involved the future occu-

Figure 12.1   Interior of Danish cohousing scheme designed by architect Jan Gudmond Hoyer in Roskilde. Here houses have been built on to a re-used portal frame factory building. The old factory provides an open communal space which can be used for parties, sport, children's play, etc. (photo: T. Woolley)

pants, and many of the innovative features could only have been introduced with the agreement of the residents.

Examples in the UK

While many of the British participatory schemes have been criticised as architecturally disappointing (Hannay, 1986), this is largely because of the exceptionally conservative attitude of financing bodies like the Housing Corporation and regulatory authorities. The architectural forms of many-user participation schemes in the UK are not as experimental as those in Denmark and other European countries. However, a closer look at the many housing cooperative and self-build schemes of the 1980s shows that the designs do reflect some tenants' preferences. In particular, the desire to create a sense of neighbourhood and bond communities together has led to clusters of housing around courtyards and other inward-looking spaces. The wide variety of layouts adopted by the different Liverpool cooperatives, for instance, is evidence of how, despite severe financial and planning controls, each scheme is quite different and pressures to standardise have been successfully overcome.

Future occupants of housing can agree to trade-offs so that savings in some areas can lead to higher standards in others, and there is evidence of this being done by housing cooperative groups. Many schemes have remained self-managed so reducing problems of maintenance and vandalism which affect many public sector schemes.

Sadly, political support for self-help and cooperative housing in the UK has waned under the Conservative government and the early promise of the cooperative movement has not been followed up. The positive benefits to those people who have taken part in such schemes are easy to identify, and projects like the Weller Streets Housing Cooperative (MacDonald, 1986) and the Eldonians in Liverpool (Cowan et al., 1988) have attracted a great deal of attention.

Participation in France and Holland

User participation is particularly widespread in France, and there are many exciting housing projects, both private cooperatives and public sector schemes. In particular some groups have experimented with self-build and the possibility of changing the internal layout of developments after they have been built. Several of these schemes are illustrated in Hatch (1984). The concept of support structures, pioneered by John Habraken (Habraken, 1972), has influenced some French schemes, and there are also many in the Netherlands, where housing associations have built in flexibility which allows family changes to be accommodated, even within flatted complexes.

The potential for even limited flexibility within housing can be of great benefit to occupants who may not have to find another house when their family size increases or decreases, or their pattern of living changes. One project outside Paris at Meudon includes play room spaces for children shared between houses and workshops for adults who work from home. None of these ideas needs to be complicated in architectural terms; the inflexibility is largely a result of narrow-minded attitudes on the part of designers and developers who can only conceive of housing in a traditional form.

Working from home

The idea of working from home is another attractive and innovative idea for housing. With new technology and changing attitudes, many people prefer to have a workspace within the home rather than commuting. It also has attractions for childcare. There are many examples of user-initiated schemes which include such flexibility, especially in Germany. One scheme in Kassel, part of the Documenta Urbana, comprises low-cost housing, finished with some help from the occupants and workshops, literally at the bottom of the garden. This allows the occupants space to work near home or extra room for hobbies which would be impossible in the low-cost minimal housing units. The scheme also exhibits a wide mix of dwelling size, including a communal group house, and many ecological features.

It was possible to initiate a similar project in Glasgow – the Workspace Housing Scheme at Paisley Road Toll, on a corner of the Garden Festival site in 1988. While occupant participation was not possible here, because of the lack of local community groups (the area had largely been cleared), the scheme offers the opportunity to work from home, high-energy efficiency and the possibility of communal spaces. An association has been formed by the residents to manage the scheme, and it serves as a model of how innovative ideas in housing can be put into practice with progressive support from local and central government officials.

Ecological housing

Many user-participation projects exhibit ecological features because, when given the opportunity, many people are concerned that their housing reflects their desire to protect natural resources and reduce the wastage of non-renewable energy. While alternative technology has been around for a long time, its application to normal housing is still in its infancy. However, user-initiated projects often have this high on their list of priorities. As a result, there are a number of schemes (see Figure 12.2), especially in Germany because of the strength of the green movement (Woolley, 1989a).

In the UK there have been a number of experiments with so-called energy saver homes, but these are distinguished by an obsession with making the houses look the same as any normal

Figure 12.2   Ecological self-build housing in Hamburg. Four houses are linked by a shared, south-facing conservatory. The building is timber-framed with much of the construction work done by the occupants (photo: T. Woolley)

developer schemes. In other European countries, however, the approach has been the opposite. In the many ecological areas in German suburbs, the new ecological houses have become propaganda for the green movement. They are easily identified by their prominent passive solar conservatories and green roofs.

Such houses are not merely concerned with saving energy, but with creating a healthy and uplifting environment for the occupants. Healthy housing emphasises the importance of quality of life and environment, which mass housing often fails to provide.

Building materials used in conventional houses contain toxic substances and these are breathed in by the occupants. While a little has been done to remove the worst examples, and reduce the use of CFCs in insulation products, there are still many concerns about the effects on the health of occupants, particularly in draught-free highly insulated houses.

Consumer control of housing gives future users the opportunity to demand healthy buildings and the application of a new design philosophy, and information on how to achieve this is growing (Pearson, 1989). If the occupants of housing take the initiative and demand what they want, housing will not be imposed in an alien way; it will grow out of the way people want to live. Architects of the future will need to respond to these aspirations and funding bodies will need to agree to finance them.

We need to be optimistic about the future and recognise that simply providing housing as numbers is no longer good enough. Much of it will be rejected and will fail as in the past. Instead, we need to ensure that housing is built to meet how we want to live. It can be done.

CONCLUSIONS

It is not possible in the space of a short chapter to indicate fully the extent of the innovative housing ideas that have now been put into practice. Anyone interested in learning more is well advised to visit the cohousing schemes of Denmark, the ecological housing projects of Germany, the participatory schemes in France or the communal housing associations of the Netherlands. Women's groups who feel instinctively that the traditional approach to housing design tends to reinforce the stereotypical image of the woman, tied to the kitchen sink looking after the children, will be amazed to discover the way in which housing design can support other ways of living.

Unfortunately, in the UK in particular, we are culturally very conservative and housing development is hemmed in with many financial, bureaucratic and attitudinal restrictions. Only by the consumers of housing asserting greater control over the design and development process through design participation are we likely to see any significant change. It is to be hoped that pressure for such change will come from women clients and women architects who will work together to produce living environments for people and children which will reflect the way we would like to live in the next century.

REFERENCES

Colquhoun, I. and Fauset, P. G. (1991) *Housing Design – An International Perspective*, London: Batsford.
Cowan, R., Hannay, P. and Owens, R. (1988) 'Community-led regeneration by the Eldonians: the light on top of the tunnel', *Architects Journal*, 23 March: 37–63.
Dalholm, H. E. (ed.) (1991) *Full-scale Modelling: Applications and Development of the Method*, Lund: University of Lund.

Dunleavy, P. (1981) *The Politics of Mass Housing in Britain 1945–1975*, Oxford: Clarendon Press.

Eno, S. and Treanor, D. (1982) *The Collective Housing Handbook*, London: Laurieston Hall Publications.

Franck, K. A. and Ahrentzen, S. (eds) (1991) *New Households, New Housing*, New York: Van Nostrand Reinhold.

Habraken, N. J. (1972) *Supports: An Alternative to Mass Housing*, London: Architectural Press. (Originally published in 1961.)

Hannay, P. (1986) 'Participation in Austria', *Architects Journal*, 3 December: 32–6.

Hatch, R. (ed.) (1984) *The Scope of Social Architecture*, New York: Van Nostrand Reinhold.

MacDonald, A. (1986) *The Weller Way*, London: Faber & Faber.

Mercer, J. (1984) *Communes – A Social History and Guide*, London: Prism Press.

Pearson, D. (1989) *The Natural House Book*, London: Gaia Books.

Pearson, L. F. (1988) *The Architectural and Social History of Co-operative Living*, Basingstoke: Macmillan.

Pepper, D. (1991) *Communes and the Green Vision*, London: Green Print.

RIBA/Institute of Housing (1988) *Tenant Participation in Housing Design: A Guide for Action*, London: RIBA Publications.

RIBA (1993) *Women Architects: A Report*, London: RIBA Publications.

Rigby, A. (1974) *Communes in Britain*, London: Routledge & Kegan Paul.

Roberts, M. (1991) *Living in a Man Made World: Gender Assumptions in Modern Housing Design*, London: Routledge.

Vale, R. and Vale, B. (1991) *Green Architecture*, London: Thames & Hudson.

Wates, N. and Wolmar, C. (eds) (1980) *Squatting, the Real Story*, London: Bay Leaf Books.

Woolley, T. (1987) '1:1 and face to face', *Architects Journal*, 29 April: 22–3.

Woolley, T. (1989a) 'Kassel – German architects and ecological settlements', *Architecture Today*, No. 5.

Woolley, T. (1989b) *Design Participation Today*. Unpublished research report for the Architects Registration Council of the UK.

Woolley, T. (1990) 'Design or rubble?', *Roof*, November/December: 25–6.

Chapter 13

An agenda for action
Issues of choice, freedom and control

Rose Gilroy

We hope that this collection of perspectives on women and housing has added to the 'where we are discourse' which seeks to address women's absence from the literature and to reassert their importance as a focus for research.

While there is much research on women and housing issues, there remains a lengthy list of unanswered questions and policy areas in which women's specific needs are not on the agenda.

It is (sadly) still relevant to draw on the structure set out by the Royal Town Planning Institute in 1989 in their influential publication *Planning for Choice and Opportunity*.

> *The majority of planning documents still do not acknowledge the importance of the gender perspective and its effect on calculation of housing need.*

The beginning of any new building or renovation process is the calculation of housing need. It is important that such calculations do not omit a gender perspective as seen in:

Demographic changes – the most obvious of which is the growing number of elderly people, the majority of whom are women. How does this group want to live? Another change is the growth in single parent households largely because of the rising divorce rate. What do single mothers want from their housing? A housing scheme built by Nomad Housing Group in Bedlington (south Northumberland) in 1990 for single parents provided a communal nursery in response to research into their life-style needs and aspirations, not just their housing needs.

A re-examination of our concept of housing need. Those familiar with local authority and housing association sector allocation systems will know the term 'adequately housed'. Consider this example. A woman who is a single parent with a young child is housed in a two-bedroom house on an unpopular run-down estate where there is a high level of burglary. This woman is frightened to be in her home and as frightened to leave it. Will she be rehoused or will she be told that she is adequately housed because her need for certain physical spaces has been met? Sophie Watson (1988) talks, as Jane Darke does, of the concept of home, which for women is associated with 'warmth, security, privacy, control and emotional relations' (Watson, 1988: 141). Does the lack of any one of these amount to housing need? Those who campaign for greater recognition and response to the issue of domestic violence would say, yes. Julia Smailes points to the real need to live near other lesbians or some focus for lesbian support. Plainly, there is a need for policy-makers to consider carefully the social and emotional aspects of housing need, rather than a simple numerical and house condition approach.

Women are disproportionately affected by unsatisfactory housing and by reduced investment in housing.

A pressing question is the definition of affordability. Roberts (1991) talks of recognising the identity of the client group in the calculation of affordability levels: 'affordability should be computed on the basis of women's earnings not the now, often outmoded, idea that there is a male wage earner and therefore more likely to be someone earning at or above the decency threshold' (Roberts, 1991: 157). Already this voice of reasonableness seems destined to be ignored. The new financial regime for housing associations, introduced in 1988, immediately led to rising rents for new build properties and in 1993 a further cut in grant rate has taken place reducing Housing Association Grant (HAG) to a 62 per cent ceiling. The implications of this third cut in HAG are depressing. New figures from Cardiff University already show that, for the first time, it is more expensive to become a tenant in a new association property than to take on a typical first-time mortgage. The average monthly mortgage (September 1993) interest payment for first-time buyers is £217.95 or £50.30 per week. Compare this with the NFHA's *CORE* bulletin for the same period which reveals average weekly rents for new dwellings as

£50.41 (Khanum, 1993). Of course, those taking on a mortgage will generally have to find deposits and have a responsibility for repair and maintenance. Nevertheless, on the plus side, interest rates do rise and fall, while rent levels only rise, and at the end of a mortgage term the householder will possibly have a capital asset. The greater concern is that this reduction of grant and the concomitant rise of rents have created a deepening housing underclass of those who can only afford their home as long as they remain out of waged work. Leeds Federated Housing Association director Peter Redman sums up the feelings of many associations: 'We must question our role in increasingly only housing people who will never be economically active, despite any recovery in the economy because our rent levels are too high to enable them to leap the poverty trap' (quoted in Khanum, 1993: 9).

How many of these households are women headed? Summaries of CORE data (NFHA, 1993) for the quarter October to December 1992 reveal that 21 per cent of allocations in this period went to single parents, and that nationally there has been a swing away from housing elderly people to housing families with children. Recent research reveals that:

> 90 % of lone parents stuck claiming benefits want to work but cannot either because the benefit and tax system makes it impossible for them to start, or because lack of childcare blocks their access to training, or because they are forced to look for work with their existing skills which often means they cannot command wages that leave them any better off after paying for childcare and work expenses.
>
> (Weston, 1993: 6)

An ideal world would see thinking about childcare as automatic as consideration of car parking. There is something seriously wrong with our system of values if spaces for the safe leaving of tin boxes are higher on the agenda than places for children. Even with attention to childcare, many women in housing association property will have no increase in disposable income because of the severity of benefit tapers. In the autumn of 1993 there were considerable fears that central government's desire to cap public spending would be borne by increasing the taper, or by reducing the ceiling on savings, or by paying recipients a maximum of 90 per cent of the amount they currently receive. The imposition of any of these changes will be to increase the hardship faced by many households – the low paid, the elderly, the single parent. In all these groups, women are in the majority.

The downturn in investment does not simply militate against intending tenants of housing association property. Central government has also made a sustained attack on resources going into the private sector, which sits oddly with their adherence to a property-owning democracy. Nationally, the specified capital grant allocation for private sector renovation has been cut by 6 per cent in 1993/94 and will be cut back more severely in the next financial year. Coupled with a cut in grant rates from 90 per cent in 1989 to 60 per cent in 1993/94 (Broadbent, 1993), this means that central government's contribution to private sector renewal is falling considerably. The cause for concern which this raises is made worse when we consider that, in the absence of more recent information, research (Nationwide, 1988) reveals that women headed households are likely to be buying poorer quality property at the bottom end of the market and in need of major investment.

A successful initiative of the last decade has been the establishment of Care and Repair schemes which help older people (and for older people we should substitute women) to make any changes necessary for their comfortable continuance in their own homes. The continuing issue for these schemes is the ability to put together a successful funding package for the work. An evaluation by Philip Leather and Sheila Mackintosh (1993) found that many of the dwellings which were tackled in the early 1980s relied, if only in part, on building society maturity loans. A problem now is that the general decline in property values, combined with the poor values of some older women's homes, may mean that societies are unwilling to lend where they see the possibility of an asset becoming a financial liability. For those in areas where crime has driven many households away, there may be no point in exploring such staying put options when a sense of community has been destroyed. For older women in these neighbourhoods a more suitable option may be sheltered housing, where again demand far outstrips supply.

As a result of their inferior economic position, women have fewer housing options.

Housing policy in this country is increasingly dominated by the marketplace. This is evident not only in the drive to owner occupation but also in escalating rents in all parts of the rented sector. The poverty trap created by rising housing association rents has already been discussed, but of equal concern is the move to capital value rents taken by some local authorities, leaving those with little or no benefit entitlement unable to take up an offer of a better quality dwelling located in a safer area. While the British Crime Survey tells us that women are less likely than men to be victims of crime, studies carried out at the estate and area level show that women are victimised by being followed, kerb crawled, stared at etc. These will not be treated as crimes by the police, yet this kind of victimisation may cause women to self-impose curfews or to go out prepared for attack (Painter, 1992).

A policy of promoting owner occupation at the expense of other tenures benefits men rather than women.

The chapter on owner occupation (Chapter 3) has already described how women are not benefiting from the housing market, or from initiatives intended to reduce purchase price or decrease mortgage payments in the early, difficult years. What initiatives would work? Perhaps those which would allow greater flexibility, such as the ability to staircase down as well as up which would prevent homelessness in hard times. Other mechanisms might be an agency (the local authority or a housing association?) which would buy property from elderly people who, through a depressed market, were unable to move to more suitable accommodation. Such flexibility would need a sea change in our housing policy, replacing a central concern for tenure with a focus on the individual and her needs.

Lending and allocation policies still discriminate against women.

It remains a remarkable truth that while a number of investigations into the allocation of rented housing have been successfully undertaken by the Commission for Racial Equality (CRE), the Equal Opportunities Commission has not been active in seeking to judge the equality of access that women may or may not enjoy. Yet, as the chapters by Perminder Dhillon-Kashyap (6) and

Julia Smailes (8), demonstrate, many women do face discrimination in allocations. A Black woman may pursue her grievance
under the Race Relations Act 1976, but what redress might a
White lesbian have in these circumstances? A number of good
practice guidelines (Dutta and Taylor, 1989; NFHA, 1992; Dean,
1993) are now available for those organisations seeking to make
radical change in their practice but, while the ethos is one of
voluntarism, only those organisations which are aware of their
responsibilities to the whole community may be making changes.
The National Federation of Housing Associations (1992) reveals
that, of the associations responding to its questionnaire on equal
opportunities issues, 83 per cent did not monitor lettings for
gender, 71 per cent did not monitor for disability and 98 per cent
did not monitor for sexuality. There are no commensurate figures
for local authorities but again research is needed to determine the
baseline if we are to measure progress.

Research, as the CRE has proved, can play a part in bringing
about change, if only because to be found wanting may create
embarrassment. A greater pressure would be for landlords who
find any portion of their funding from central government to be
asked to make public their performance on issues of equality
(Gilroy, 1992). At present, for local authorities Circular 19/90
states more than 40 indicators on which authorities must publish
information. In addition, there is flexibility to include other matters which the authority considers of importance. Why should
issues of equality not be included on the list of mandatory matters?
For example:

- descriptions of, and statistics relating to, policies on harassment (because of race, gender, sexuality);
- a description of the monitoring system used to track allocations and nominations to housing associations with a breakdown by race and gender (and including disability and
  sexuality);
- information on a rolling review of the examination of housing
  services undertaken with the need to highlight and eradicate
  discriminatory practice;
- descriptions of methods used to determine housing need
  among different groups.

In the field of lending for owner occupation, again the CRE has
been active in bringing to light poor practice (1985, 1988) on the

part of estate agents and building societies. These investigations were carried out long after the Race Relations Act made their practices unlawful. Compare the position regarding women. There is a widespread view that women are no longer discriminated against by lenders because of the passing of the Sex Discrimination Act 1975, but is this true or simply an assumption? At present, the drive to owner occupation is being made at the expense of invest-ment in rented housing which disadvantages women because fewer women can meet the requirements for owner occupation. A case of indirect discrimination? If so, one which would never make any progress.

Of course, the British obsession with owner occupation is bound up with the Government's thoughts about the family structure. How are we to resolve the tension between the nuclear family-centred owner occupation drive and the reality of British life: the rising divorce rate, rising homeless figures and a million house-holds with negative equity?

Women suffer disproportionately from homelessness.

Because women's housing needs often arise out of sudden life events – pregnancy, relationship breakdown, sexual or physical violence – they are more likely to be found among the homeless than men. Official statistics based on a view of homelessness dictated by central government show women as mothers making up the largest group of homeless households. Pat Niner's research (1989) revealed that women fleeing a violent partner were often only given priority status if they had children; in short, that we are only valued as mothers. For those without children who have neither friends nor family to take them in nor financial resources to find independent accommodation, the only option is to stay and suffer. More emergency accommodation is needed with adequate revenue support to maintain quality and level of support.

Greve and Currie (1990) talk of the high risk of homelessness faced by those with low incomes, particularly those who are unem-ployed or single parents or from ethnic minority groups. Consider how the numbers made vulnerable may increase because of rising rents and possible changes to Housing Benefit. How many may fall into arrears and lose their homes? Consider the ominous attacks on young single mothers by members of central government. The absurd concept that young women become pregnant to jump the housing queue ahead of respectable young couples waiting for housing may be just playing to the media, or it may result in pregnancy among the unmarried and the under-25s being declared an act of intentionality. Where will these women go? Will we see the wave of young single people sleeping on the streets followed by a wave of young mothers? Will social services departments step in under the Children Act 1989, and would such intervention simply mean more children in care?

In respect of the single homeless, Greve and Currie list among other 'disadvantages' the fact of being female. The size of the homeless problem among women is likely to be under-estimated. The Census figures on rough sleepers reveal the likelihood of undercounting and the problem that women's homelessness, unlike that of men, is private not public. The result of our hidden homelessness is that there is far less hostel provision for women. Women are less likely to apply for places in mixed hostels for fear of harassment and because the facilities, regime and management style are male centred. Again, more research and more financial attention are desperately needed.

Women are affected disproportionately by insensitive housing design which generally takes inadequate account of those under-taking caring and domestic work at home. A number of principles are not given adequate consideration in private and public sector developments.

In considerations of housing design, women are still seen as house-wives first and foremost (Roberts, 1991). As the chapters by Karen Franck (11) and Tom Woolley (12) reveal, a recognition that women are workers, whether at home or outside, and re-search into how individuals want to live enrich the architectural process. In this country the cautious views of those funding owner occupation have led to a stifling of imaginative solutions. Greater flexibility has been demonstrated by the local authority and the housing association sector which has embraced concepts such as 'housing for life' and 'visitability'. We still need to grasp the reality that many people work at home and therefore need space to do so. This might mean working at a computer or it might mean packing Christmas cards in boxes or stitching waistbands on skirts. This is the less attractive side of home working, yet for many women it is their only opportunity to add to their income and look after their children. For Black women there is the added problem of racial discrimination in the job market and harassment in the workplace, all of which may keep a woman within her own four walls.

Is there still scope for creative thinking about design? To what extent has innovation been hampered by new funding regimes which have pushed many housing associations to design and build packages from developers? Are there possibilities for growth in the numbers of women entering self-build and deciding for themselves what housing they want? In Britain self-build is seen as male but, as Caroline Moser and Linda Peake (1987) discuss, women in other countries are in the forefront of such initiatives, so why not here?

Our examination of women and housing issues has not been limited to issues of access and design. We have also considered the place of housing education and, leading on from that, the barriers to increasing the number of women in positions of power to improve quality of life for other women. In all areas of education there is a deep-rooted problem that the majority of educators are White middle-class men and educational resources are produced by the same group. This one vision of the world then becomes encoded as knowledge which denies the possibility of other visions. Consider this indictment of planning education: 'In town planning education, students go through four years' full-time study and in this period they are fed an educational diet of implicit and explicit racist and sexist assumptions' (Grey and Amooquaye, 1990: 232).

Research is needed into the content and style of *housing* education before we may say that this is not applicable. Similarly, the addressing of equal opportunities issues is not limited to course content but needs to include issues such as the timing of classes, childcare, and help with fees for those with little financial resource. It is vital that all these issues are given priority attention during validation inspections. As Marion Brion tells us (Chapter 9), there are now greater numbers of women studying for housing qualifications than ever before. However, as Veronica Coatham and Janet Hale's chapter (10) reveals, the number finding their way to the top is, if anything, decreasing in number.

There are general issues which act as barriers to women in employment, such as the response to caring responsibilities. Many women are faced with the impossible decision of never having a child, or of having a child and working full time while paying out all their earnings in private childcare. Others elect to work part time including job share, and see their career progression halted because fewer hours are seen as less of a commitment to the job (Coyle, 1989); others have a child, work and engage in rigid timetabling with family members. These remain choices that women make, not men. Of course, apart from the pre-birth and birth stages, there is no reason why child-rearing should be seen as exclusively or even primarily a women's issue. However, until men, perhaps through the impact of divorce and the Children Act rather than a change in values, have to cope with childcare issues, little progress is likely to be made. Similarly, the growing numbers of old people and the move to so-called community care may mean that a woman passes from caring for her children to caring for her elderly parents to caring for her sick partner. The impact of any of these may be that a woman has to curtail her own development and her career with subsequent impact on her independence, her financial standing and her self-esteem. A fundamental change is needed to see that women are not natural (and therefore happy to be unpaid) carers. This might lead to the political will to unlock resources to support dependent adults and those who care for their welfare.

These are common barriers which all women workers find erected against them, but what of housing work itself? Are we reverting back to a more macho image? Only a few years ago the new emphasis on customer care and landlord/tenant relationships seemed to suggest that women might come more to the fore. The most recent change in a field of public policy constantly buffeted by change is the move to Compulsory Competitive Tendering. Are we entering a world of creeping hard managerialism which is seen as a world of men (Coyle 1989)? Ongoing research is needed to determine how opportunities for women are increased or eroded by policy changes.

A depressing picture perhaps. The effects of changes in housing policy in the last decade have been especially felt by women. Our dependence on the male wage for our entry into owner occupation makes us vulnerable to homelessness when relationship break-down occurs. Our reliance on the rented sector makes us vulnerable when government has turned its face away. We continue to suffer from invisibility on so many fronts that we must end by a call for more research on women in all tenures and in the world of housing work. Without facts to draw attention to our oppressed position we will continue to find it difficult to lobby for change.

REFERENCES

Broadbent, H. (1993) 'We can rebuild it', *Housing*, July: 50–1.

Commission for Racial Equality (1985) *Race and Mortgage Lending: Formal Investigation of Mortgage Lending in Rochdale*, London: CRE Publications.

Commission for Racial Equality (1988) *Racial Discrimination in a London Estate Agency. Report of a Formal Investigation into Richard Barclay & Co*, London: CRE Publications.

Coyle, A. (1989) 'The limits of change in local government and equal opportunities for women', *Public Administration*, Vol 67, Spring: 39–50.

Dean, R. (1993) *Policy into Practice: A Guide to Equal Opportunity Action Plans in Housing Services*, Coventry: Institute of Housing.

Dutta, R. and Taylor, G (1989) *Housing Equality: An Action Guide*, London: CHAR.

Gilroy, R. (1992) 'Taking equality on board', *Housing*, February: 38.

Greve, J. and Currie, E. (1990) *Homelessness in Britain*, York: Joseph Rowntree Foundation.

Grey, G. and Amooquaye, E. (1990) 'A new agenda for race and planning', in J. Montgomery and A. Thornley (eds) *Radical Planning Initiatives*, Aldershot: Gower.

Khanum, S. (1993) 'On the brink', *Inside Housing*, 17 September: 8–9.

Leather, P. and Mackintosh, S. (1993) 'Staying power', *Housing*, July: 53–4.

Moser, C. O. N. and Peake, L. (1987) *Women, Human Settlements and Housing*, London: Tavistock.

National Federation of Housing Associations (1992) *Equal Opportunities in Housing Associations: Are You Doing Enough?*, London: NFHA.

National Federation of Housing Associations (1993) *CORE Quarterly Bulletin*, No 13, October–December 1992.

Nationwide Anglia Building Society (1988) *Lending to Women*.

Niner, P. (1989) *Homelessness in Nine Local Authorities: Case Studies of Policy and Practice*, London: HMSO.

Painter, K. (1992) 'Different worlds', in D. Evans, N. R. Fyfe and D. T. Herbert (eds) *Crime, Policing and Place: Essays in Environmental Criminology*, London: Routledge.

Roberts, M. (1991) *Living in a Man Made World: Gender Assumptions in Modern Housing Design*, London: Routledge.

Royal Town Planning Institute (1989) *Planning for Choice and Opportunity*, London: RTPI.

Watson, S. (1988) *Accommodating Inequality: Gender and Housing*, Sydney: Allen & Unwin.

Weston, C. (1993) 'Women, discrimination and work', *AUT Bulletin*, October: 6–7.

Index

Nomad Housing Group,
  Northumberland 260
Northumberland: harassment in
  accommodation 135; HAY 131,
  144–6, 147; homelessness 127,
  132; Nomad Housing Group 260
'not rich, not poor' research 80, 81
NTO (National Tenants
  Organisation) 64
nuclear family 127, 153, 226–8

Oakley, Ann 17, 23
occupational status, and tenure
  39, 40, 41
older women 5–6; disability 88;
  disadvantaged 95–6; housing
  78–9, 89–92. 94–8; living alone
  77–8; pensions 84–7; property
  values 263; statistics 76–7;
  vulnerability 75
Oldham estate agent, CRE 104,
  118
opportunity, ladder of 191–2; *see
  also* equal opportunities
Opportunity 2000 177
owner occupation 31–6; caring
  daughter 42–5; disadvantages
  51, 53; and divorce or single
  parents 36–42; encouraged 264;
  expense 35; geographical
  differences 47; government
  White Paper 35; low-cost 103–4;
  marriage 32–3; and nuclear
  family 35; purchase schemes
  53–4; quality 55; shared 53;
  women's participation 3, 33–4,
  36, 42, 49–54, 55, 264

Painter, K. 264
parents, custodial status 36, 38
participatory housing 252, 254–5
Peach, C. 49, 116
Peake, Linda 58–9, 268
Pearson, D. 257–8
Pearson, L. 251
pensioners: *see* elderly people
pensions 44–5, 82, 84–7
Planning for Choice and

Opportunity (Royal Town
  Planning Institute) 260
Plater-Zyberk, ELizabeth 243
Positive Action Training in
  Housing schemes 185
poverty, defined 93
*Poverty in Black and White –
  Deprivation and the Ethnic
  Minorities* 109
poverty trap 141–2, 264
pregnancy, and homelessness
  138–40, 267
Priority Estates Projects 61
private renting 33, 45–6, 119, 142
private sector: grant cuts 263;
  house buying schemes 53–4;
  property values 263; *see also*
  owner occupation
Proven, B. 203–4
public housing *see* council housing

Race Relations Act (1976) 108,
  121–2, 265, 266
racial discrimination 104
*Racial Discrimination in Hostel
  Accommodation* (CRE) 104
racial harassment 105–7
racism: institutional 105, 112–13;
  structural 105–12; subjective
  112–13
Rajan, A. 199, 211
Rao, Nirmala 101, 107, 110, 111,
  112, 113, 115–16
Redman, Peter 262
renting: affordability 142–4;
  allocation of housing 53, 264–5;
  fair/assured 142, 144; legislation
  109; local authority 34; and
  mortgage payments, compared
  261–2; NFHA 261; private
  sector 33, 45–6, 119, 142;
  unemployed women 45–6; *see
  also* council housing
repairs, housing 68
Right to Buy 49–53
Ritzdorf, M. 227
Roberts, Marion 18, 19–20, 251,
  261, 267
Roof (Shelter) 248